Teacher, Student, and Parent One-Stop Internet Resources

Log on to bookm.msscience.com

ONLINE STUDY TOOLS
- Section Self-Check Quizzes
- Interactive Tutor
- Chapter Review Tests
- Standardized Test Practice
- Vocabulary PuzzleMaker

ONLINE RESEARCH
- WebQuest Projects
- Prescreened Web Links
- Career Links
- Internet Labs

INTERACTIVE ONLINE STUDENT EDITION
- Complete Interactive Student Edition available at mhln.com

FOR TEACHERS
- Teacher Bulletin Board
- Teaching Today—Professional Development

SAFETY SYMBOLS

SAFETY SYMBOLS	HAZARD	EXAMPLES	PRECAUTION	REMEDY
DISPOSAL	Special disposal procedures need to be followed.	certain chemicals, living organisms	Do not dispose of these materials in the sink or trash can.	Dispose of wastes as directed by your teacher.
BIOLOGICAL	Organisms or other biological materials that might be harmful to humans	bacteria, fungi, blood, unpreserved tissues, plant materials	Avoid skin contact with these materials. Wear mask or gloves.	Notify your teacher if you suspect contact with material. Wash hands thoroughly.
EXTREME TEMPERATURE	Objects that can burn skin by being too cold or too hot	boiling liquids, hot plates, dry ice, liquid nitrogen	Use proper protection when handling.	Go to your teacher for first aid.
SHARP OBJECT	Use of tools or glassware that can easily puncture or slice skin	razor blades, pins, scalpels, pointed tools, dissecting probes, broken glass	Practice common-sense behavior and follow guidelines for use of the tool.	Go to your teacher for first aid.
FUME	Possible danger to respiratory tract from fumes	ammonia, acetone, nail polish remover, heated sulfur, moth balls	Make sure there is good ventilation. Never smell fumes directly. Wear a mask.	Leave foul area and notify your teacher immediately.
ELECTRICAL	Possible danger from electrical shock or burn	improper grounding, liquid spills, short circuits, exposed wires	Double-check setup with teacher. Check condition of wires and apparatus.	Do not attempt to fix electrical problems. Notify your teacher immediately.
IRRITANT	Substances that can irritate the skin or mucous membranes of the respiratory tract	pollen, moth balls, steel wool, fiberglass, potassium permanganate	Wear dust mask and gloves. Practice extra care when handling these materials.	Go to your teacher for first aid.
CHEMICAL	Chemicals can react with and destroy tissue and other materials	bleaches such as hydrogen peroxide; acids such as sulfuric acid, hydrochloric acid; bases such as ammonia, sodium hydroxide	Wear goggles, gloves, and an apron.	Immediately flush the affected area with water and notify your teacher.
TOXIC	Substance may be poisonous if touched, inhaled, or swallowed.	mercury, many metal compounds, iodine, poinsettia plant parts	Follow your teacher's instructions.	Always wash hands thoroughly after use. Go to your teacher for first aid.
FLAMMABLE	Flammable chemicals may be ignited by open flame, spark, or exposed heat.	alcohol, kerosene, potassium permanganate	Avoid open flames and heat when using flammable chemicals.	Notify your teacher immediately. Use fire safety equipment if applicable.
OPEN FLAME	Open flame in use, may cause fire.	hair, clothing, paper, synthetic materials	Tie back hair and loose clothing. Follow teacher's instruction on lighting and extinguishing flames.	Notify your teacher immediately. Use fire safety equipment if applicable.

 Eye Safety Proper eye protection should be worn at all times by anyone performing or observing science activities.

 Clothing Protection This symbol appears when substances could stain or burn clothing.

 Animal Safety This symbol appears when safety of animals and students must be ensured.

 Handwashing After the lab, wash hands with soap and water before removing goggles.

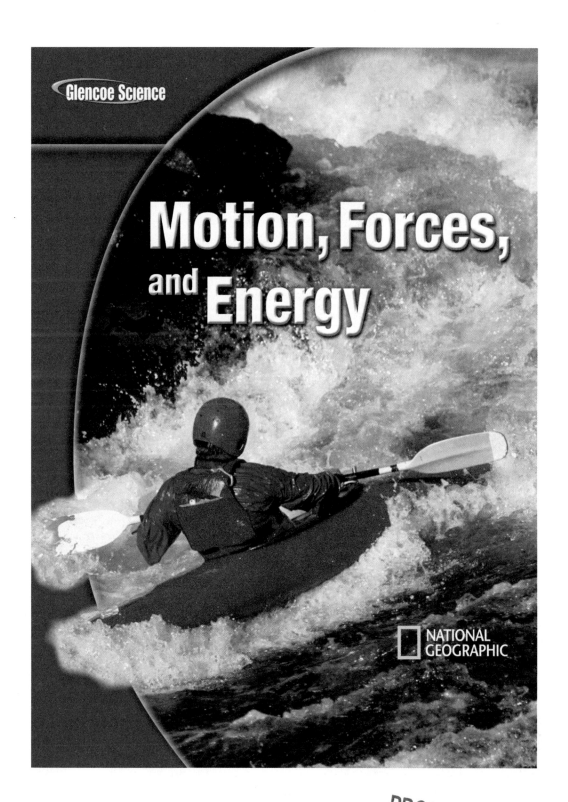

Motion, Forces, and Energy

Glencoe Science

NATIONAL GEOGRAPHIC

New York, New York Columbus, Ohio Chicago, Illinois Woodland Hills, California

Glencoe Science

Motion, Forces, and Energy

This kayaker battles the rapids on the Thompson River in British Columbia, Canada. A kayaker takes advantage of Newton's third law. The paddle exerts a force on the water and the water exerts an equal, but opposite, force on the kayaker.

The McGraw·Hill Companies

Copyright © 2008 by The McGraw-Hill Companies, Inc. All rights reserved. Except as permitted under the United States Copyright Act, no part of this publication may be reproduced or distributed in any form or by any means, or stored in a database or retrieval system, without prior written permission of the publisher.

The National Geographic features were designed and developed by the National Geographic Society's Education Division. Copyright © National Geographic Society. The name "National Geographic Society" and the Yellow Border Rectangle are trademarks of the Society, and their use, without prior written permission, is strictly prohibited.

The "Science and Society" and the "Science and History" features that appear in this book were designed and developed by TIME School Publishing, a division of TIME Magazine. TIME and the red border are trademarks of Time Inc. All rights reserved.

Send all inquiries to:
Glencoe/McGraw-Hill
8787 Orion Place
Columbus, OH 43240-4027

ISBN: 978-0-07-877836-0
MHID: 0-07-877836-0

Printed in the United States of America.

1 2 3 4 5 6 7 8 9 10 027/043 09 08 07

Authors

NATIONAL GEOGRAPHIC
Education Division
Washington, D.C.

Thomas McCarthy, PhD
Science Department Chair
St. Edward's School
Vero Beach, FL

Dinah Zike
Educational Consultant
Dinah-Might Activities, Inc.
San Antonio, TX

Deborah Lillie
Math and Science Writer
Sudbury, MA

Margaret K. Zorn
Science Writer
Yorktown, VA

Series Consultants

CONTENT

Jack Cooper
Ennis High School
Ennis, TX

Sandra K. Enger, PhD
Associate Director, Associate Professor
UAH Institute for Science Education
Huntsville, AL

Carl Zorn, PhD
Staff Scientist
Jefferson Laboratory
Newport News, VA

MATH

Michael Hopper, DEng
Manager of Aircraft Certification
L-3 Communications
Greenville, TX

Teri Willard, EdD
Mathematics Curriculum Writer
Belgrade, MT

READING

Barry Barto
Special Education Teacher
John F. Kennedy Elementary
Manistee, MI

SAFETY

Aileen Duc, PhD
Science 8 Teacher
Hendrick Middle School, Plano ISD
Plano, TX

Sandra West, PhD
Department of Biology
Texas State University-San Marcos
San Marcos, TX

ACTIVITY TESTERS

Nerma Coats Henderson
Pickerington Lakeview Jr. High School
Pickerington, OH

Mary Helen Mariscal-Cholka
William D. Slider Middle School
El Paso, TX

Science Kit and Boreal Laboratories
Tonawanda, NY

Series Reviewers

Sharla Adams
IPC Teacher
Allen High School
Allen, TX

John Barry
Seeger Jr- Sr. High School
West Lebanon, IN

Nora M. Prestinari Burchett
Saint Luke School
McLean, VA

Joanne Davis
Murphy High School
Murphy, NC

Sandra Everhart
Dauphin/Enterprise Jr. High Schools
Enterprise, AL

Lynne Huskey
Chase Middle School
Forest City, NC

Michelle Mazeika-Simmons
Whiting Middle School
Whiting, IN

Mark Sailer
Pioneer Jr-Sr. High School
Royal Center, IN

HOW TO...
Use Your Science Book

Why do I need my science book?

Have you ever been in class and not understood all of what was presented? Or, you understood everything in class, but at home, got stuck on how to answer a question? Maybe you just wondered when you were ever going to use this stuff?

These next few pages are designed to help you understand everything your science book can be used for . . . besides a paperweight!

Before You Read

- **Chapter Opener** Science is occurring all around you, and the opening photo of each chapter will preview the science you will be learning about. The **Chapter Preview** will give you an idea of what you will be learning about, and you can try the **Launch Lab** to help get your brain headed in the right direction. The **Foldables** exercise is a fun way to keep you organized.

- **Section Opener** Chapters are divided into two to four sections. The **As You Read** in the margin of the first page of each section will let you know what is most important in the section. It is divided into four parts. **What You'll Learn** will tell you the major topics you will be covering. **Why It's Important** will remind you why you are studying this in the first place! The **Review Vocabulary** word is a word you already know, either from your science studies or your prior knowledge. The **New Vocabulary** words are words that you need to learn to understand this section. These words will be in **boldfaced** print and highlighted in the section. Make a note to yourself to recognize these words as you are reading the section.

As You Read

- **Headings** Each section has a title in large red letters, and is further divided into blue titles and small red titles at the beginnings of some paragraphs. To help you study, make an outline of the headings and subheadings.

- **Margins** In the margins of your text, you will find many helpful resources. The **Science Online** exercises and **Integrate** activities help you explore the topics you are studying. **MiniLabs** reinforce the science concepts you have learned.

- **Building Skills** You also will find an **Applying Math** or **Applying Science** activity in each chapter. This gives you extra practice using your new knowledge, and helps prepare you for standardized tests.

- **Student Resources** At the end of the book you will find **Student Resources** to help you throughout your studies. These include **Science, Technology,** and **Math Skill Handbooks,** an **English/Spanish Glossary,** and an **Index.** Also, use your **Foldables** as a resource. It will help you organize information, and review before a test.

- **In Class** Remember, you can always ask your teacher to explain anything you don't understand.

FOLDABLES Study Organizer

Science Vocabulary Make the following Foldable to help you understand the vocabulary terms in this chapter.

STEP 1 Fold a vertical sheet of notebook paper from side to side.

STEP 2 Cut along every third line of only the top layer to form tabs.

STEP 3 Label each tab with a vocabulary word from the chapter.

Build Vocabulary As you read the chapter, list the vocabulary words on the tabs. As you learn the definitions, write them under the tab for each vocabulary word.

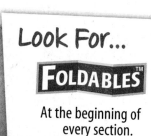

Look For...
FOLDABLES
At the beginning of every section.

In Lab

Working in the laboratory is one of the best ways to understand the concepts you are studying. Your book will be your guide through your laboratory experiences, and help you begin to think like a scientist. In it, you not only will find the steps necessary to follow the investigations, but you also will find helpful tips to make the most of your time.

- Each lab provides you with a **Real-World Question** to remind you that science is something you use every day, not just in class. This may lead to many more questions about how things happen in your world.

- Remember, experiments do not always produce the result you expect. Scientists have made many discoveries based on investigations with unexpected results. You can try the experiment again to make sure your results were accurate, or perhaps form a new hypothesis to test.

- Keeping a **Science Journal** is how scientists keep accurate records of observations and data. In your journal, you also can write any questions that may arise during your investigation. This is a great method of reminding yourself to find the answers later.

Look For...

- **Launch Labs** start every chapter.
- **MiniLabs** in the margin of each chapter.
- **Two Full-Period Labs** in every chapter.
- **EXTRA Try at Home Labs** at the end of your book.
- the **Web site** with **laboratory demonstrations**.

Before a Test

Admit it! You don't like to take tests! However, there *are* ways to review that make them less painful. Your book will help you be more successful taking tests if you use the resources provided to you.

- Review all of the **New Vocabulary** words and be sure you understand their definitions.
- Review the notes you've taken on your **Foldables,** in class, and in lab. Write down any question that you still need answered.
- Review the **Summaries** and **Self Check questions** at the end of each section.
- Study the concepts presented in the chapter by reading the **Study Guide** and answering the questions in the **Chapter Review.**

Look For...
- **Reading Checks** and **caption questions** throughout the text.
- the **Summaries** and **Self Check questions** at the end of each section.
- the **Study Guide** and **Review** at the end of each chapter.
- the **Standardized Test Practice** after each chapter.

Let's Get Started

To help you find the information you need quickly, use the Scavenger Hunt below to learn where things are located in Chapter 1.

1. What is the title of this chapter?
2. What will you learn in Section 1?
3. Sometimes you may ask, "Why am I learning this?" State a reason why the concepts from Section 2 are important.
4. What is the main topic presented in Section 2?
5. How many reading checks are in Section 1?
6. What is the Web address where you can find extra information?
7. What is the main heading above the sixth paragraph in Section 2?
8. There is an integration with another subject mentioned in one of the margins of the chapter. What subject is it?
9. List the new vocabulary words presented in Section 2.
10. List the safety symbols presented in the first Lab.
11. Where would you find a Self Check to be sure you understand the section?
12. Suppose you're doing the Self Check and you have a question about concept mapping. Where could you find help?
13. On what pages are the Chapter Study Guide and Chapter Review?
14. Look in the Table of Contents to find out on which page Section 2 of the chapter begins.
15. You complete the Chapter Review to study for your chapter test. Where could you find another quiz for more practice?

Teacher Advisory Board

The Teacher Advisory Board gave the editorial staff and design team feedback on the content and design of the Student Edition. They provided valuable input in the development of the 2008 edition of *Glencoe Science.*

John Gonzales
Challenger Middle School
Tucson, AZ

Rachel Shively
Aptakisic Jr. High School
Buffalo Grove, IL

Roger Pratt
Manistique High School
Manistique, MI

Kirtina Hile
Northmor Jr. High/High School
Galion, OH

Marie Renner
Diley Middle School
Pickerington, OH

Nelson Farrier
Hamlin Middle School
Springfield, OR

Jeff Remington
Palmyra Middle School
Palmyra, PA

Erin Peters
Williamsburg Middle School
Arlington, VA

Rubidel Peoples
Meacham Middle School
Fort Worth, TX

Kristi Ramsey
Navasota Jr. High School
Navasota, TX

Student Advisory Board

The Student Advisory Board gave the editorial staff and design team feedback on the design of the Student Edition. We thank these students for their hard work and creative suggestions in making the 2008 edition of *Glencoe Science* student friendly.

Jack Andrews
Reynoldsburg Jr. High School
Reynoldsburg, OH

Peter Arnold
Hastings Middle School
Upper Arlington, OH

Emily Barbe
Perry Middle School
Worthington, OH

Kirsty Bateman
Hilliard Heritage Middle School
Hilliard, OH

Andre Brown
Spanish Emersion Academy
Columbus, OH

Chris Dundon
Heritage Middle School
Westerville, OH

Ryan Manafee
Monroe Middle School
Columbus, OH

Addison Owen
Davis Middle School
Dublin, OH

Teriana Patrick
Eastmoor Middle School
Columbus, OH

Ashley Ruz
Karrer Middle School
Dublin, OH

The Glencoe middle school science Student Advisory Board taking a timeout at COSI, a science museum in Columbus, Ohio.

Contents

Nature of Science: Science in Motion—2

Chapter 1 — Motion and Momentum—6
- **Section 1** What is motion? 8
- **Section 2** Acceleration 14
- **Section 3** Momentum 19
  - Lab Collisions 25
  - Lab: Design Your Own
    Car Safety Testing 26

Chapter 2 — Force and Newton's Laws—34
- **Section 1** Newton's First Law 36
- **Section 2** Newton's Second Law 42
- **Section 3** Newton's Third Law 49
  - Lab Balloon Races 55
  - Lab: Design Your Own
    Modeling Motion in Two Directions 56

Chapter 3 — Forces and Fluids—64
- **Section 1** Pressure 66
- **Section 2** Why do objects float? 74
  - Lab Measuring Buoyant Force 81
- **Section 3** Doing Work with Fluids 82
  - Lab: Use the Internet
    Barometric Pressure and Weather 88

Chapter 4 — Work and Simple Machines—96
- **Section 1** Work and Power 98
  - Lab Building the Pyramids 103
- **Section 2** Using Machines 104
- **Section 3** Simple Machines 109
  - Lab: Design Your Own Pulley Power 116

In each chapter, look for these opportunities for review and assessment:
- Reading Checks
- Caption Questions
- Section Review
- Chapter Study Guide
- Chapter Review
- Standardized Test Practice
- Online practice at bookm.mssymbol.com

Get Ready to Read Strategies
- Summarize 8A
- Compare and Contrast 36A
- Make Inferences 66A
- Questions and Answers 98A
- Take Notes 126A
- Identify the Main Idea 158A

Contents

Chapter 5: Energy and Energy Resources—124

- **Section 1** What is energy? 126
- **Section 2** Energy Transformations 131
  - Lab Hearing with Your Jaw 138
- **Section 3** Sources of Energy 139
  - Lab: Use the Internet
    Energy to Power Your Life 148

Chapter 6: Thermal Energy—156

- **Section 1** Temperature and Thermal Energy 158
- **Section 2** Heat 162
  - Lab Heating Up and Cooling Down 168
- **Section 3** Engines and Refrigerators 169
  - Lab: Design Your Own
    Comparing Thermal Insulators 174

Student Resources

Science Skill Handbook—184
- Scientific Methods 184
- Safety Symbols 193
- Safety in the Science
  Laboratory 194

Extra Try at Home Labs—196

Technology Skill Handbook—199
- Computer Skills 199
- Presentation Skills 202

Math Skill Handbook—203
- Math Review 203
- Science Applications 213

Reference Handbooks—218
- Physical Science
  Reference Tables 218
- Periodic Table of
  the Elements 220

English/Spanish Glossary—222

Index—228

Cedits—233

M ◆ xi

Cross-Curricular Readings/Labs

available as a video lab

NATIONAL GEOGRAPHIC VISUALIZING

1 The Conservation
   of Momentum 23
2 Newton's Laws in Sports 51
3 Pressure at Varying
   Temperatures 72
4 Levers 113
5 Energy Transformations 134
6 The Four-Stroke Cycle 171

TIME SCIENCE AND Society

2 Air Bag Safety 58
4 Bionic People 118
6 The Heat is On 175

Oops! Accidents in SCIENCE

1 What Goes Around
   Comes Around 28

Science and Language Arts

3 "Hurricane" 90

SCIENCE Stats

5 Energy to Burn 150

Launch LAB

1 Motion After a Collision 7
2 Forces and Motion 35
3 Forces Exerted by Air 65
4 Compare Forces 97
5 Marbles and Energy 125
6 Measuring Temperature 157

Mini LAB

1 Modeling Acceleration 17
2 Measuring Force Pairs 53
3 Interpreting Footprints 68
4 Observing Pulleys 114
5 Building a Solar Collector 143
6 Observing Convection 165

Mini LAB Try at Home

1 Measuring Average Speed 11
2 Observing Friction 40
3 Observing Bernoulli's Principle . . . 85
4 Work and Power 101
5 Analyzing Energy
   Transformations 133
6 Comparing Rates of Melting . . . 164

One-Page Labs

1 Collisions 25
2 Balloon Races 55
3 Measuring Buoyant Force 81
4 Building the Pyramids 103
5 Hearing with Your Jaw 138
6 Heating Up and
   Cooling Down 168

Labs/Activities

Design Your Own Labs

- **1** Car Safety Testing 26–27
- **2** Modeling Motion in Two Directions 56–57
- **4** Pulley Power 116–117
- **6** Comparing Thermal Insulators 174–175

Use the Internet Labs

- **3** Barometric Pressure and Weather 88–89
- **5** Energy to Power Your Life 148–149

Applying Math

- **1** Speed of a Swimmer 10
- **1** Acceleration of a Bus 16
- **1** Momentum of a Bicycle 20
- **2** Acceleration of a Car 45
- **3** Calculating Pressure 67
- **4** Calculating Work 100
- **4** Calculating Power 101
- **4** Calculating Mechanical Advantage 105
- **4** Calculating Efficiency 107
- **6** Converting to Celsius 160

Applying Science

- **3** Layering Liquids 78
- **5** Is energy consumption outpacing production? 142

Astronomy: 69
Career: 78, 172
Earth Science: 140
History: 43, 100
Life Science: 10, 37, 50, 87, 107, 111, 133, 135, 166, 167
Physics: 90
Social Studies: 20

Science Online

12, 22, 39, 50, 67, 84, 102, 105, 132, 142, 170

Standardized Test Practice

32–33, 62–63, 94–95, 122–123, 154–155, 180–181

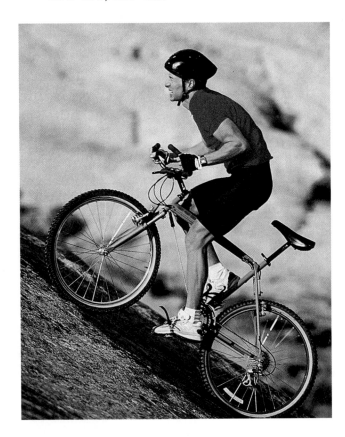

M ◆ 1

The NATURE OF SCIENCE

Scientific Methods

Science in Motion

How do scientists learn more about the world? Scientists usually follow an organized set of procedures to solve problems. These procedures are called scientific methods. Although the steps in these methods can vary depending on the type of problem a scientist is solving, they all are an organized way of asking a question, forming a possible answer, investigating the answer, and drawing conclusions about the answer. Humans have investigated questions about motion for thousands of years, asking questions such as: "What causes motion? How fast do things fall? How does a pendulum work?" However, scientific methods were not always used to learn the answers to these questions.

Figure 1 Motion, like on this busy road, is all around you.

Early Scientists

The ancient Greeks believed in supernatural beings—gods and goddesses—whose powers made the world work. In the 500s B.C., a group of Greek philosophers in the city of Miletus proposed that natural events should only be explained by what humans can learn with their senses—sight, hearing, smell, touch, and taste. 200 years later, Aristotle, a Greek philosopher and teacher, developed a system of logic for distinguishing truth from falsehood. He also studied plants and animals and recorded detailed observations of them. Because of these practices, he is considered one of the first scientists as they are defined today.

Figure 2 At his school in ancient Athens, Greece, Aristotle taught philosophy and science.

Aristotle also investigated how objects move and why they continue moving. He believed that the speed of a falling object depends on its weight. Unfortunately, Aristotle did not have a scientific method to test his ideas. Therefore, Aristotle's untested theory was not proven wrong for hundreds of years. In the late 1500s, Italian scientist Galileo Galilei conducted experiments to test Aristotle's ideas. He rolled balls down inclined planes and swung pendulums, measuring how far and how fast they moved. According to legend, Galileo climbed to the top of the Tower of Pisa and dropped two objects of different weights. They hit the ground at the same time, finally proving that the speed of a falling object doesn't depend on its weight.

Figure 3 Galileo made observations of the motion of pendulums in order to learn about motion.

Developing Scientific Methods

Galileo and others that came after him gradually developed new ideas about how to learn about the universe. These new methods of scientific investigation were different from the methods used by earlier philosophers in an important way. A scientific explanation makes predictions that can be tested by observations of the world or by doing experiments. If the predictions are not supported by the observations or experiments, the scientific explanation cannot be true and has to be changed or discarded.

Figure 4 A scientific explanation of motion would explain the motion of this roller coaster.

The Nature of Science

Physical Science

The study of motion, forces, and energy is part of physical science. Physical scientists also learn about elements, atoms, electricity, sound, and more.

Like all scientists, they use experimentation and careful observation to answer questions about how the world works. Other scientists learn about these experiments and try to repeat them. In this way, scientists eliminate the flaws in their work and participate in the search for answers.

Scientific Methods

The understanding of motion was undertaken by philosophers such as Descartes and scientists such as Galileo. Their efforts led to the creation of procedures, called scientific methods, which scientists use to investigate the world. Scientific methods generally include several steps.

Identifying a Question

The first step in a scientific method is to identify a question to be answered. For example, Aristotle wanted to know what causes motion. The answer to one question often leads to others. Aristotle wondered how an object's weight affects the speed at which it falls. After Galileo showed that an object's weight does not affect its falling speed, Newton wanted to know how fast objects fall, regardless of their weight.

Forming a Hypothesis

The next step is to form a hypothesis. A hypothesis is a possible answer to the question that is consistent with available information. A hypothesis can result from analyzing data or from observations. For example, data show that lung cancer occurs more frequently in smokers than in nonsmokers. A hypothesis might be that smoking causes lung cancer. Observations of falling objects might lead to the hypothesis that heavier objects fall faster than lighter ones.

Scientific Methods

1. **Identify a question.**
   Determine a question to be answered.
2. **Form a hypothesis.**
   Gather information and propose an answer to the question.
3. **Test the hypothesis.**
   Perform experiments or make observations to see if the hypothesis is supported.
4. **Analyze results.**
   Look for patterns in the data that have been collected.
5. **Draw a conclusion.**
   Decide what the test results mean. Communicate your results.

Scientists use scientific methods to answer questions about motion.

Testing a Hypothesis

A hypothesis must be testable to see if it is correct. This is done by performing experiments and measuring the results. Galileo tested Aristotle's hypothesis by rolling balls of differing weight down an inclined plane to see which, if either, rolled faster. Since a well-designed experiment is crucial, Galileo made sure the inclined plane was smooth and the balls were released in the same way.

Analyzing Results

Scientists collect information, called data, which must be analyzed. In order to organize, study, and detect patterns in data, scientists use graphs and other methods.

Collecting data requires careful measurements. Many experiments of the past were flawed because the measuring devices were inaccurate. Because Galileo needed precise timing, he used a water clock to measure the time for a ball to roll down the inclined plane. If his clock had been inaccurate, Galileo's results would have been less useful.

Drawing a Conclusion

The last step in a scientific experiment is to draw a conclusion based on results and observations. Sometimes the data does not support the original hypothesis and scientists must start the process again, beginning with a new hypothesis. Other times, though, the data supports the original hypothesis. If a hypothesis is supported by repeated experiments, it can become a theory—an idea that has withstood repeated testing and is used to explain observations. Scientists, however, know that nothing is certain. A new idea, a new hypothesis, and a new experiment can alter what is believed to be true about the world.

Figure 5 These students are conducting an experiment to learn how objects move.

You Do It

A ball may fall, but will it bounce back? What determines how high and how fast it will bounce? Make a list of possible factors that affect the way a ball bounces. Choose one of these and form a hypothesis about it. Think of experiments you could do to test your hypothesis.

chapter 1

Motion and Momentum

The BIG Idea

The motion of an object can be described by its velocity.

SECTION 1
What is motion?
Main Idea Motion is a change in position.

SECTION 2
Acceleration
Main Idea Acceleration occurs when an object speeds up, slows down, or changes direction.

SECTION 3
Momentum
Main Idea In a collision, momentum can be transferred from one object to another.

Springing into Action

The hunt might just be over for this pouncing leopard. A leopard can run as fast as 60 km/h over short distances and can jump as high as 3 m. However, a leopard must be more than fast and strong. To catch its prey, it must also be able to change its speed and direction quickly.

Science Journal Describe how your motion changed as you moved from your school's entrance to your classroom.

Start-Up Activities

Motion After a Collision

How is it possible for a 70-kg football player to knock down a 110-kg football player? The smaller player usually must be running faster. Mass makes a difference when two objects collide, but the speed of the objects also matters. Explore the behavior of colliding objects during this lab.

1. Space yourself about 2 m away from a partner. Slowly roll a baseball on the floor toward your partner, and have your partner roll a baseball quickly into your ball.
2. Have your partner slowly roll a baseball as you quickly roll a tennis ball into the baseball.
3. Roll two tennis balls toward each other at the same speed.
4. **Think Critically** In your Science Journal, describe how the motion of the balls changed after the collisions, including the effects of speed and type of ball.

 Preview this chapter's content and activities at bookm.msscience.com

 Motion and Momentum
Make the following Foldable to help you understand the vocabulary terms in this chapter.

STEP 1 Fold a vertical sheet of notebook paper from side to side.

STEP 2 Cut along every third line of only the top layer to form tabs.

STEP 3 Label each tab.

Build Vocabulary As you read the chapter, list the vocabulary words about motion and momentum on the tabs. As you learn the definitions, write them under the tab for each vocabulary word.

M ◆ 7

Get Ready to Read

Summarize

① Learn It! Summarizing helps you organize information, focus on main ideas, and reduce the amount of information to remember. To summarize, restate the important facts in a short sentence or paragraph. Be brief and do not include too many details.

② Practice It! Read the text on page 21 labeled *Conservation of Momentum*. Then read the summary below and look at the important facts from that passage.

Important Facts

- A collision doesn't change the total momentum of a group of objects.
- Friction is an outside force that causes the total momentum to decrease.
- Objects can bounce off each other when they collide.
- Objects can stick together when they collide.

Summary

The total momentum of a group of objects remains constant unless an outside force acts on the objects.

③ Apply It! Practice summarizing as you read this chapter. Stop after each section and write a brief summary.

Target Your Reading

Reading Tip
Reread your summary to make sure you didn't change the author's original meaning or ideas.

Use this to focus on the main ideas as you read the chapter.

1 Before you read the chapter, respond to the statements below on your worksheet or on a numbered sheet of paper.
- Write an **A** if you **agree** with the statement.
- Write a **D** if you **disagree** with the statement.

2 After you read the chapter, look back to this page to see if you've changed your mind about any of the statements.
- If any of your answers changed, explain why.
- Change any false statements into true statements.
- Use your revised statements as a study guide.

Science Online
Print out a worksheet of this page at bookm.msscience.com

Before You Read A or D		Statement	After You Read A or D
	1	Distance traveled and displacement are always equal.	
	2	When an object changes direction, it is accelerating.	
	3	A horizontal line on a distance-time graph means the speed is zero.	
	4	If two objects are moving at the same speed, the heavier object is harder to stop.	
	5	The instantaneous speed of an object is always equal to its average speed.	
	6	Momentum equals mass divided by velocity.	
	7	Speed always is measured in kilometers per hour.	
	8	An object's momentum increases if its speed increases.	
	9	If a car is accelerating, its speed must be increasing.	
	10	Speed and velocity are the same thing.	

section 1

What is motion?

as you read

What You'll Learn
- **Define** distance, speed, and velocity.
- **Graph** motion.

Why It's Important
The different motions of objects you see every day can be described in the same way.

Review Vocabulary
meter: SI unit of distance, abbreviated m; equal to 39.37 in

New Vocabulary
- speed
- average speed
- instantaneous speed
- velocity

Matter and Motion

All matter in the universe is constantly in motion, from the revolution of Earth around the Sun to electrons moving around the nucleus of an atom. Leaves rustle in the wind. Lava flows from a volcano. Bees move from flower to flower as they gather pollen. Blood circulates through your body. These are all examples of matter in motion. How can the motion of these different objects be described?

Changing Position

To describe an object in motion, you must first recognize that the object is in motion. Something is in motion if it is changing position. It could be a fast-moving airplane, a leaf swirling in the wind, or water trickling from a hose. Even your school, attached to Earth, is moving through space. When an object moves from one location to another, it is changing position. The runners shown in **Figure 1** sprint from the start line to the finish line. Their positions change, so they are in motion.

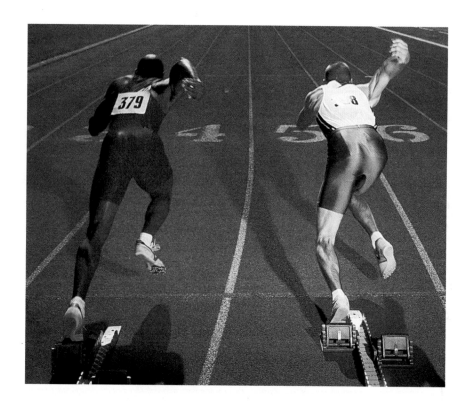

Figure 1 These sprinters are in motion because their positions change.

8 ◆ **M**   **CHAPTER 1**   Motion and Momentum

Relative Motion Determining whether something changes position requires a point of reference. An object changes position if it moves relative to a reference point. To visualize this, picture yourself competing in a 100-m dash. You begin just behind the start line. When you pass the finish line, you are 100 m from the start line. If the start line is your reference point, then your position has changed by 100 m relative to the start line, and motion has occurred. Look at **Figure 2.** How can you determine that the dog has been in motion?

 How do you know if an object has changed position?

Distance and Displacement Suppose you are to meet your friends at the park in five minutes. Can you get there on time by walking, or should you ride your bike? To help you decide, you need to know the distance you will travel to get to the park. This distance is the length of the route you will travel from your house to the park.

Suppose the distance you traveled from your house to the park was 200 m. When you get to the park, how would you describe your location? You could say that your location was 200 m from your house. However, your final position depends on both the distance you travel and the direction. Did you go 200 m east or west? To describe your final position exactly, you also would have to tell the direction from your starting point. To do this, you would specify your displacement. Displacement includes the distance between the starting and ending points and the direction in which you travel. **Figure 3** shows the difference between distance and displacement.

Figure 2 Motion occurs when something changes position relative to a reference point.
Explain *how the position of the dog changed.*

Figure 3 Distance is how far you have walked. Displacement is the direction and difference in position between your starting and ending points.

Distance: 40 m
Displacement: 40 m east

Distance: 70 m
Displacement: 50 m northeast

Distance: 140 m
Displacement: 0 m

SECTION 1 What is motion? **M ◆ 9**

Animal Speeds Different animals can move at different top speeds. What are some of the fastest animals? Research the characteristics that help animals run, swim, or fly at high speed.

Speed

To describe motion, you usually want to describe how fast something is moving. The faster something is moving, the greater the distance it can travel in a unit of time, such as one second or one hour. **Speed** is the distance an object travels in a unit of time. For example, an object with a speed of 5 m/s can travel 5 m in 1 s. Speed can be calculated from this equation:

Speed Equation

$$\text{speed (in meters/second)} = \frac{\text{distance (in meters)}}{\text{time (in seconds)}}$$

$$s = \frac{d}{t}$$

The unit for speed is the unit of distance divided by the unit of time. In SI units, speed is measured in units of m/s—meters per second. However, speed can be calculated using other units such as kilometers for distance and hours for time.

Applying Math — Solve a Simple Equation

SPEED OF A SWIMMER Calculate the speed of a swimmer who swims 100 m in 56 s.

Solution

1 *This is what you know:*
- distance: $d = 100$ m
- time: $t = 56$ s

2 *This is what you need to know:*
speed: $s = ?$ m/s

3 *This is the procedure you need to use:*
Substitute the known values for distance and time into the speed equation and calculate the speed:
$$s = \frac{d}{t} = \frac{100 \text{ m}}{56 \text{ s}} = \frac{100}{56} \frac{\text{m}}{\text{s}} = 1.8 \text{ m/s}$$

4 *Check your answer:*
Multiply your answer by the time. You should get the distance that was given.

Practice Problems

1. A runner completes a 400-m race in 43.9 s. In a 100-m race, he finishes in 10.4 s. In which race was his speed faster?
2. A passenger train travels from Boston to New York, a distance of 350 km, in 3.5 h. What is the train's speed?

 For more practice, visit bookm.msscience.com/math_practice

10 ◆ M CHAPTER 1 Motion and Momentum

Average Speed A car traveling in city traffic might have to speed up and slow down many times. How could you describe the speed of an object whose speed is changing? One way is to determine the object's average speed between where it starts and stops. The speed equation on the previous page can be used to calculate the average speed. **Average speed** is found by dividing the total distance traveled by the total time taken.

Reading Check *How is average speed calculated?*

Instantaneous Speed An object in motion can change speeds many times as it speeds up or slows down. The speed of an object at one instant of time is the object's **instantaneous speed.** To understand the difference between average and instantaneous speeds, think about walking to the library. If it takes you 0.5 h to walk 2 km to the library, your average speed would be as follows:

$$s = \frac{d}{t}$$
$$= \frac{2 \text{ km}}{0.5 \text{ h}} = 4 \text{ km/h}$$

However, you might not have been moving at the same speed throughout the trip. At a crosswalk, your instantaneous speed might have been 0 km/h. If you raced across the street, your speed might have been 7 km/h. If you were able to walk at a steady rate of 4 km/h during the entire trip, you would have moved at a constant speed. Average speed, instantaneous speed, and constant speed are illustrated in **Figure 4.**

Measuring Average Speed

Procedure
1. Choose two points, such as two doorways, and mark each with a small piece of **masking tape.**
2. Measure the distance between the two points.
3. Use a **watch, clock,** or **timer** that indicates seconds to time yourself walking from one mark to the other.
4. Time yourself walking slowly, walking safely and quickly, and walking with a varying speed; for example, slow/fast/slow.

Analysis
1. Calculate your average speed in each case.
2. Predict how long it would take you to walk 100 m slowly, at your normal speed, and quickly.

Try at Home

Figure 4 The average speed of each ball is the same from 0 s to 4 s.

The top ball is moving at a constant speed. In each second, the ball moves the same distance.

The bottom ball has a varying speed. Its instantaneous speed is fastest between 0 s and 1 s, slower between 2 s and 3 s, and slowest between 3 s and 4 s.

Graphing Motion

You can represent the motion of an object with a distance-time graph. For this type of graph, time is plotted on the horizontal axis and distance is plotted on the vertical axis. **Figure 5** shows the motion of two students who walked across a classroom plotted on a distance-time graph.

Distance-Time Graphs and Speed A distance-time graph can be used to compare the speeds of objects. Look at the graph shown in **Figure 5**. According to the graph, after 1 s student A traveled 1 m. Her average speed during the first second is as follows:

$$\text{speed} = \frac{\text{distance}}{\text{time}} = \frac{1 \text{ m}}{1 \text{ s}} = 1 \text{ m/s}$$

Student B, however, traveled only 0.5 m in the first second. His average speed is

$$\text{speed} = \frac{\text{distance}}{\text{time}} = \frac{0.5 \text{ m}}{1 \text{ s}} = 0.5 \text{ m/s}$$

So student A traveled faster than student B. Now compare the steepness of the lines on the graph in **Figure 5**. The line representing the motion of student A is steeper than the line for student B. A steeper line on the distance-time graph represents a greater speed. A horizontal line on the distance-time graph means that no change in position occurs. In that case, the speed of the object is zero.

Topic: Land Speed Record
Visit bookm.msscience.com for Web links to information about how the land speed record has changed over the past century.

Activity Make a graph showing the increase in the land speed over time.

Figure 5 The motion of two students walking across a classroom is plotted on this distance-time graph.
Use the graph to determine which student had the faster average speed.

Velocity

The motion of an object also depends on the direction in which the object is moving. The direction of an object's motion can be described with its velocity. The **velocity** of an object is the speed of the object and the direction of its motion. For example, if a car is moving west with a speed of 80 km/h, the car's velocity is 80 km/h west. The velocity of an object is sometimes represented by an arrow. The arrow points in the direction in which the object is moving. **Figure 6,** uses arrows to show the velocities of two hikers.

The velocity of an object can change if the object's speed changes, its direction of motion changes, or they both change. For example, suppose a car is traveling at a speed of 40 km/h north and then turns left at an intersection and continues on with a speed of 40 km/h. The speed of the car is constant at 40 km/h, but the velocity changes from 40 km/h north to 40 km/h west. Why can you say the velocity of a car changes as it comes to a stop at an intersection?

Figure 6 The arrows show the velocities of two hikers. Although the hikers have the same speed, they have different velocities because they are moving in different directions.

section 1 review

Summary

Changing Position
- An object is in motion if it changes position relative to a reference point.
- Motion can be described by distance, speed, displacement, and velocity, where displacement and velocity also include direction.

Speed and Velocity
- The speed of an object can be calculated by dividing the distance traveled by the time needed to travel the distance.
- For an object traveling at constant speed, its average speed is the same as its instantaneous speed.
- The velocity of an object is the speed of the object and its direction of motion.

Graphing Motion
- A line on a distance-time graph becomes steeper as an object's speed increases.

Self Check

1. **Identify** the two pieces of information you need to know the velocity of an object.
2. **Make and Use Graphs** You walk forward at 1.5 m/s for 8 s. Your friend decides to walk faster and starts out at 2.0 m/s for the first 4 s. Then she slows down and walks forward at 1.0 m/s for the next 4 s. Make a distance-time graph of your motion and your friend's motion. Who walked farther?
3. **Think Critically** A bee flies 25 m north of the hive, then 10 m east, 5 m west, and 10 m south. How far north and east of the hive is it now? Explain how you calculated your answer.

Applying Math

4. **Calculate** the average velocity of a dancer who moves 5 m toward the left of the stage over the course of 15 s.
5. **Calculate Travel Time** An airplane flew a distance of 650 km at an average speed of 300 km/h. How much time did the flight take?

section 2

Acceleration

as you read

What You'll Learn
- **Define** acceleration.
- **Predict** what effect acceleration will have on motion.

Why It's Important
Whenever the motion of an object changes, it is accelerating.

Review Vocabulary
kilogram: SI unit of mass, abbreviated kg; equal to approximately 2.2 lbs

New Vocabulary
- acceleration

Acceleration and Motion

When you watch the first few seconds of a liftoff, a rocket barely seems to move. With each passing second, however, you can see it move faster until it reaches an enormous speed. How could you describe the change in the rocket's motion? When an object changes its motion, it is accelerating. **Acceleration** is the change in velocity divided by the time it takes for the change to occur.

Like velocity, acceleration has a direction. If an object speeds up, the acceleration is in the direction that the object is moving. If an object slows down, the acceleration is opposite to the direction that the object is moving. What if the direction of the acceleration is at an angle to the direction of motion? Then the direction of motion will turn toward the direction of the acceleration.

Speeding Up You get on a bicycle and begin to pedal. The bike moves slowly at first, but speeds up as you keep pedaling. Recall that the velocity of an object is the speed of an object and its direction of motion. Acceleration occurs whenever the velocity of an object changes. Because the bike's speed is increasing, the velocity of the bike is changing. As a result, the bike is accelerating.

For example, the toy car in **Figure 7** is accelerating because it is speeding up. The speed of the car is 10 cm/s after 1s, 20 cm/s after 2s, and 30 cm/s after 3s. Here the direction of the car's acceleration is in the same direction as the car's velocity—to the right.

Figure 7 The toy car is accelerating because its speed is increasing.

14 ◆ **M CHAPTER 1** Motion and Momentum

Slowing Down Now suppose you are biking at a speed of 4 m/s and you apply the brakes. This causes you to slow down. When you slow down, your velocity changes because your speed decreases. This means that acceleration occurs when an object slows down, as well as when it speeds up. The car in **Figure 8** is slowing down. During each time interval, the car travels a smaller distance, so its speed is decreasing.

In both of these examples, speed is changing, so acceleration is occurring. Because speed is decreasing in the second example, the direction of the acceleration is opposite to the direction of motion. Any time an object slows down, its acceleration is in the direction opposite to the direction of its motion.

Changing Direction The velocity of an object also changes if the direction of motion changes. Then the object doesn't move in a straight line, but instead moves in a curved path. The object is accelerating because its velocity is changing. In this case the direction of acceleration is at an angle to the direction of motion.

Again imagine yourself riding a bicycle. When you turn the handlebars, the bike turns. Because the direction of the bike's motion has changed, the bike has accelerated. The acceleration is in the direction that the bicycle turned.

Figure 9 shows another example of an object that is accelerating. The ball starts moving upward, but its direction of motion changes as its path turns downward. Here the acceleration is downward. The longer the ball accelerates, the more its path turns toward the direction of acceleration.

Reading Check *What are three ways to accelerate?*

Figure 8 The car is moving to the right but accelerating to the left. In each time interval, it covers less distance and moves more slowly.
Determine *how the car's velocity is changing.*

Figure 9 The ball starts out by moving forward and upward, but the acceleration is downward, so the ball's path turns in that direction.

SECTION 2 Acceleration **M** ◆ **15**

Calculating Acceleration

If an object is moving in only one direction, its acceleration can be calculated using this equation.

Acceleration Equation

acceleration (in m/s²) =
$$\frac{(\text{final speed (in m/s)} - \text{initial speed (in m/s)})}{\text{time (in s)}}$$

$$a = \frac{(s_f - s_i)}{t}$$

In this equation, time is the length of time over which the motion changes. In SI units, acceleration has units of meters per second squared (m/s²).

Applying Math — Solve a Simple Equation

ACCELERATION OF A BUS Calculate the acceleration of a bus whose speed changes from 6 m/s to 12 m/s over a period of 3 s.

Solution

1 *This is what you know:*
- initial speed: s_i = 6 m/s
- final speed: s_f = 12 m/s
- time: t = 3 s

2 *This is what you need to know:* acceleration: a = ? m/s²

3 *This is the procedure you need to use:* Substitute the known values of initial speed, final speed and time in the acceleration equation and calculate the acceleration:

$$a = \frac{(s_f - s_i)}{t} = \frac{(12 \text{ m/s} - 6 \text{ m/s})}{3 \text{ s}} = 6\frac{\text{m}}{\text{s}} \times \frac{1}{3 \text{ s}} = 2 \text{ m/s}^2$$

4 *Check your answer:* Multiply the calculated acceleration by the known time. Then add the known initial speed. You should get the final speed that was given.

Practice Problems

1. Find the acceleration of a train whose speed increases from 7 m/s to 17 m/s in 120 s.
2. A bicycle accelerates from rest to 6 m/s in 2 s. What is the bicycle's acceleration?

For more practice, visit bookm.msscience.com/math_practice

Figure 10 When skidding to a stop, you are slowing down. This means you have a negative acceleration.

Positive and Negative Acceleration An object is accelerating when it speeds up, and the acceleration is in the same direction as the motion. An object also is accelerating when it slows down, but the acceleration is in the direction opposite to the motion, such as the bicycle in **Figure 10.** How else is acceleration different when an object is speeding up and slowing down?

Suppose you were riding your bicycle in a straight line and increased your speed from 4 m/s to 6 m/s in 5 s. You could calculate your acceleration from the equation on the previous page.

$$a = \frac{(s_f - s_i)}{t}$$
$$= \frac{(6 \text{ m/s} - 4 \text{ m/s})}{5 \text{ s}} = \frac{+2 \text{ m/s}}{5 \text{ s}}$$
$$= +0.4 \text{ m/s}^2$$

When you speed up, your final speed always will be greater than your initial speed. So subtracting your initial speed from your final speed gives a positive number. As a result, your acceleration is positive when you are speeding up.

Suppose you slow down from a speed of 4 m/s to 2 m/s in 5 s. Now the final speed is less than the initial speed. You could calculate your acceleration as follows:

$$a = \frac{(s_f - s_i)}{t}$$
$$= \frac{(2 \text{ m/s} - 4 \text{ m/s})}{5 \text{ s}} = \frac{-2 \text{ m/s}}{5 \text{ s}}$$
$$= -0.4 \text{ m/s}^2$$

Because your final speed is less than your initial speed, your acceleration is negative when you slow down.

Modeling Acceleration

Procedure
1. Use **masking tape** to lay a course on the floor. Mark a starting point and place marks along a straight path at 10 cm, 40 cm, 90 cm, 160 cm, and 250 cm from the start.
2. Clap a steady beat. On the first beat, the person walking the course should be at the starting point. On the second beat, the walker should be on the first mark, and so on.

Analysis
1. Describe what happens to your speed as you move along the course. Infer what would happen if the course were extended farther.
2. Repeat step 2, starting at the other end. Are you still accelerating? Explain.

Figure 11 A speed-time graph can be used to find acceleration. When the line rises, the object is speeding up. When the line falls, the object is slowing down.
Infer what acceleration a horizontal line represents.

Graphing Accelerated Motion The motion of an object that is moving in a single direction can be shown with a graph. For this type of graph, speed is plotted on the vertical axis and time on the horizontal axis. Take a look at **Figure 11**. On section A of the graph, the speed increases from 0 m/s to 10 m/s during the first 2 s, so the acceleration is +5 m/s². The line in section A slopes upward to the right. An object that is speeding up will have a line on a speed-time graph that slopes upward.

Now look at section C. Between 4 s and 6 s, the object slows down from 10 m/s to 4 m/s. The acceleration is −3 m/s². On the speed-time graph, the line in section C is sloping downward to the right. An object that is slowing down will have a line on a speed-time graph that slopes downward.

On section B, where the line is horizontal, the change in speed is zero. So a horizontal line on the speed-time graph represents an acceleration of zero or constant speed.

section 2 review

Summary

Acceleration and Motion
- Acceleration is the change in velocity divided by the time it takes to make the change. Acceleration has direction.
- Acceleration occurs whenever an object speeds up, slows down, or changes direction.

Calculating Acceleration
- For motion in a single direction, acceleration can be calculated from this equation:

$$a = \frac{s_f - s_i}{t}$$

- If an object is speeding up, its acceleration is positive; if an object is slowing down, its acceleration is negative.
- On a speed-time graph, a line sloping upward represents increasing speed, a line sloping downward represents decreasing speed, and a horizontal line represents zero acceleration or constant speed.

Self Check

1. **Compare and contrast** speed, velocity, and acceleration.
2. **Infer** the motion of a car whose speed-time graph shows a horizontal line, followed by a straight line that slopes downward to the bottom of the graph.
3. **Think Critically** You start to roll backward down a hill on your bike, so you use the brakes to stop your motion. In what direction did you accelerate?

Applying Math

4. **Calculate** the acceleration of a runner who accelerates from 0 m/s to 3 m/s in 12 s.
5. **Calculate Speed** An object falls with an acceleration of 9.8 m/s². What is its speed after 2 s?
6. **Make and Use a Graph** A sprinter had the following speeds at different times during a race: 0 m/s at 0 s, 4 m/s at 2 s, 7 m/s at 4 s, 10 m/s at 6 s, 12 m/s at 8 s, and 10 m/s at 10 s. Plot these data on a speed-time graph. During what time intervals is the acceleration positive? Negative? Is the acceleration ever zero?

section 3
Collisions

A collision occurs when a moving object collides with other objects. What happens when a cue ball collides with another ball in a game of pool? The answer is that the velocities of the two balls change. The collision can change the speed of each ball, the direction of motion of each ball, or both. When a collision occurs, changes in motion of the colliding objects depend on their masses and their velocities before the collision.

Mass and Inertia

The mass of an object affects how easy it is to change its motion. **Mass** is the amount of matter in an obejct. Imagine a person rushing toward you. To stop the person, you would have to push on him or her. However, you would have to push harder to stop an adult than to stop a child. The child would be easier to stop because it has less mass than the adult. The more mass an object has, the harder it is to change its motion. In **Figure 12,** the tennis ball has more mass than the table-tennis ball. A big racquet rather than a small paddle is used to change its motion. The tendency of an object to resist a change in its motion is called **inertia**. The amount of resistance to a change in motion increases as an object's mass increases.

Reading Check *What is inertia?*

as you read

What You'll Learn
- **Define** momentum.
- **Explain** why momentum might not be conserved after a collision.
- **Predict** motion using the law of conservation of momentum.

Why It's Important
Objects in motion have momentum. The motion of objects after they collide depends on their momentum.

Review Vocabulary
triple-beam balance: scientific instrument used to measure mass precisely by comparing the mass of a sample to known masses

New Vocabulary
- mass
- inertia
- momentum
- law of conservation of momentum

Figure 12 A tennis ball has more mass than a table-tennis ball. The tennis ball must be hit harder than the table-tennis ball to change its velocity by the same amount.

SECTION 3 Momentum **M ◆ 19**

Forensics and Momentum
Forensic investigations of accidents and crimes often involve determining the momentum of an object. For example, the law of conservation of momentum sometimes is used to reconstruct the motion of vehicles involved in a collision. Research other ways momentum is used in forensic investigations.

Momentum

You know that the faster a bicycle moves, the harder it is to stop. Just as increasing the mass of an object makes it harder to stop, so does increasing the speed or velocity of the object. The **momentum** of an object is a measure of how hard it is to stop the object, and it depends on the object's mass and velocity. Momentum is usually symbolized by p.

Momentum Equation
momentum (in kg · m/s) = **mass** (in kg) × **velocity** (in m/s)
$$p = mv$$

Mass is measured in kilograms and velocity has units of meters per second, so momentum has units of kilograms multiplied by meters per second (kg · m/s). Also, because velocity includes a direction, momentum has a direction that is the same as the direction of the velocity.

 Reading Check Explain how an object's momentum changes as its velocity changes.

Applying Math — Solve a Simple Equation

MOMENTUM OF A BICYCLE Calculate the momentum of a 14-kg bicycle traveling north at 2 m/s.

Solution

1 *This is what you know:*
- mass: $m = 14$ kg
- velocity: $v = 2$ m/s north

2 *This is what you need to find:*
momentum: $p = ?$ kg · m/s

3 *This is the procedure you need to use:*
Substitute the known values of mass and velocity into the momentum equation and calculate the momentum:
$p = mv = (14$ kg$)(2$ m/s north$) = 28$ kg · m/s north

4 *Check your answer:*
Divide the calculated momentum by the mass of the bicycle. You should get the velocity that was given.

Practice Problems

1. A 10,000-kg train is traveling east at 15 m/s. Calculate the momentum of the train.
2. What is the momentum of a car with a mass of 900 kg traveling north at 27 m/s?

 For more practice, visit bookm.msscience.com/math_practice

CHAPTER 1 Motion and Momentum

Conservation of Momentum

If you've ever played billiards, you know that when the cue ball hits another ball, the motions of both balls change. The cue ball slows down and may change direction, so its momentum decreases. Meanwhile, the other ball starts moving, so its momentum increases.

In any collision, momentum is transferred from one object to another. Think about the collision between two billiard balls. If the momentum lost by one ball equals the momentum gained by the other ball, then the total amount of momentum doesn't change. When the total momentum of a group of objects doesn't change, momentum is conserved.

The Law of Conservation of Momentum According to the **law of conservation of momentum,** the total momentum of a group of objects remains constant unless outside forces act on the group. The moving cue ball and the other billiard balls in **Figure 13** are a group of objects. The law of conservation of momentum means that collisions between these objects don't change the total momentum of all the objects in the group.

Only an outside force, such as friction between the billiard balls and the table, can change the total momentum of the group of objects. Friction will cause the billiard balls to slow down as they roll on the table and the total momentum will decrease.

Types of Collisions Objects can collide with each other in different ways. **Figure 14** shows two examples. Sometimes objects will bounce off each other like the bowling ball and bowling pins. In other collisions, objects will collide and stick to each other, as when one football player tackles another.

Figure 13 When the cue ball hits the other billiard balls, it slows down because it transfers some of its momentum to the other billiard balls.
Predict *what would happen to the speed of the cue ball if all of its momentum were transferred to the other billiard balls.*

Figure 14 When objects collide, they can bounce off each other, or they can stick to each other.

When the bowling ball hits the pins, the ball and the pins bounce off each other. The momentum of the ball and the pins changes during the collision.

When one player tackles the other, they stick together. The momentum of each player changes during the collision.

Before the student on skates and the backpack collide, she is not moving.

After the collision, the student and the backpack move together at a slower speed than the backpack had before the collision.

Figure 15 Momentum is transferred from the backpack to the student.

Using Momentum Conservation The law of momentum conservation can be used to predict the velocity of objects after they collide. To use the law of conservation of momentum, assume that the total momentum of the colliding objects doesn't change.

For example, imagine being on skates when someone throws a backpack to you, as in **Figure 15.** The law of conservation of momentum can be used to find your velocity after you catch the backpack. Suppose a 2-kg backpack initially has a velocity of 5 m/s east. Your mass is 48 kg, and initially you're at rest. Then the total initial momentum is:

total momentum = momentum of backpack + your momentum
= 2 kg × 5 m/s east + 48 kg × 0 m/s
= 10 kg · m/s east

After the collision, the total momentum remains the same, and only one object is moving. Its mass is the sum of your mass and the mass of the backpack. You can use the equation for momentum to find the final velocity

total momentum = (mass of backpack + your mass) × velocity
10 kg · m/s east = (2 kg + 48 kg) × velocity
10 kg · m/s east = (50 kg) × velocity
0.2 m/s east = velocity

This is your velocity right after you catch the backpack. The velocity of you and the backpack together is much smaller than the initial velocity of the backpack. **Figure 16** shows the results of collisions between two objects that don't stick to each other.

Topic: Collisions
Visit bookm.msscience.com for Web links to information about collisions between objects with different masses.

Activity Draw diagrams showing the results of collisions between a bowling ball and a tennis ball if they are moving in the same direction and if they are in opposite directions.

NATIONAL GEOGRAPHIC VISUALIZING CONSERVATION OF MOMENTUM

Figure 16

The law of conservation of momentum can be used to predict the results of collisions between different objects, whether they are subatomic particles smashing into each other at enormous speeds, or the collisions of marbles, as shown on this page. What happens when one marble hits another marble initially at rest? The results of the collisions depend on the masses of the marbles.

A Here, a less massive marble strikes a more massive marble that is at rest. After the collision, the smaller marble bounces off in the opposite direction. The larger marble moves in the same direction that the small marble was initially moving.

B Here, the large marble strikes the small marble that is at rest. After the collision, both marbles move in the same direction. The less massive marble always moves faster than the more massive one.

C If two objects of the same mass moving at the same speed collide head-on, they will rebound and move with the same speed in the opposite direction. The total momentum is zero before and after the collision.

SECTION 3 Momentum M ◆ 23

Figure 17 When bumper cars collide, they bounce off each other, and momentum is transferred.

Colliding and Bouncing Off In some collisions, the objects involved, like the bumper cars in **Figure 17,** bounce off each other. The law of conservation of momentum can be used to determine how these objects move after they collide.

For example, suppose two identical objects moving with the same speed collide head on and bounce off. Before the collision, the momentum of each object is the same, but in opposite directions. So the total momentum before the collision is zero. If momentum is conserved, the total momentum after the collision must be zero also. This means that the two objects must move in opposite directions with the same speed after the collision. Then the total momentum once again is zero.

section 3 review

Summary

Inertia and Momentum

- Inertia is the tendency of an object to resist a change in motion.
- The momentum of an object in motion is related to how hard it is to stop the object, and can be calculated from the following equation:
  $$p = mv$$
- The momentum of an object is in the same direction as its velocity.

The Law of Conservation of Momentum

- According to the law of conservation of momentum, the total momentum of a group of objects remains constant unless outside forces act on the group.
- When objects collide, they can bounce off each other or stick together.

Self Check

1. **Explain** how momentum is transferred when a golfer hits a ball with a golf club.
2. **Determine** if the momentum of an object moving in a circular path at constant speed is constant.
3. **Explain** why the momentum of a billiard ball rolling on a billiard table changes.
4. **Think Critically** Two identical balls move directly toward each other with equal speeds. How will the balls move if they collide and stick together?

Applying Math

5. **Momentum** What is the momentum of a 0.1-kg mass moving with a velocity of 5 m/s west?
6. **Momentum Conservation** A 1-kg ball moving at 3 m/s east strikes a 2-kg ball and stops. If the 2-kg ball was initially at rest, find its velocity after the collision.

Collisions

A collision occurs when a baseball bat hits a baseball or a tennis racket hits a tennis ball. What would happen if you hit a baseball with a table-tennis paddle or a table-tennis ball with a baseball bat? How do the masses of colliding objects change the results of collisions?

Real-World Question

How does changing the size and number of objects in a collision affect the collision?

Goals

- **Compare and contrast** different collisions.
- **Determine** how the velocities after a collision depend on the masses of the colliding objects.

Materials

small marbles (5)   metersticks (2)
large marbles (2)   tape

Safety Precautions

Procedure

1. Tape the metersticks next to each other, slightly farther apart than the width of the large marbles. This limits the motion of the marbles to nearly a straight line.
2. Place a small target marble in the center of the track formed by the metersticks. Place another small marble at one end of the track. Flick the small marble toward the target marble. Describe the collision.
3. Repeat step 2, replacing the two small marbles with the two large marbles.
4. Repeat step 2, replacing the small shooter marble with a large marble.
5. Repeat step 2, replacing the small target marble with a large marble.
6. Repeat step 2, replacing the small target marble with four small marbles that are touching.
7. Place two small marbles at opposite ends of the metersticks. Shoot the marbles toward each other and describe the collision.
8. Place two large marbles at opposite ends of the metersticks. Shoot the marbles toward each other and describe the collision.
9. Place a small marble and a large marble at opposite ends of the metersticks. Shoot the marbles toward each other and describe the collision.

Conclude and Apply

1. **Describe** In which collisions did the shooter marble change direction? How did the mass of the target marble compare with the mass of the shooter marble in these collisions?
2. **Describe** how the velocity of the shooter marble changed when the target marble had the same mass and was at rest.

Communicating Your Data

Make a chart showing your results. You might want to make before-and-after sketches, with short arrows to show slow movement and long arrows to show fast movement.

LAB **M** ◆ 25

LAB Design Your Own

Car Safety Testing

Goals
- **Construct** a fast car.
- **Design** a safe car that will protect a plastic egg from the effects of inertia when the car crashes.

Possible Materials
insulated foam meat trays or fast food trays
insulated foam cups
straws, narrow and wide
straight pins
tape
plastic eggs

Safety Precautions

WARNING: *Protect your eyes from possible flying objects.*

Real-World Question
Imagine that you are a car designer. How can you create an attractive, fast car that is safe? When a car crashes, the passengers have inertia that can keep them moving. How can you protect the passengers from stops caused by sudden, head-on impacts?

Form a Hypothesis
Develop a hypothesis about how to design a car to deliver a plastic egg quickly and safely through a race course and a crash at the end.

Test Your Hypothesis

Make a Plan

1. Be sure your group has agreed on the hypothesis statement.
2. **Sketch** the design for your car. List the materials you will need. Remember that to make the car move smoothly, narrow straws will have to fit into the wider straws.

26 ◆ **M CHAPTER 1** Motion and Momentum

Using Scientific Methods

3. As a group, make a detailed list of the steps you will take to test your hypothesis.
4. Gather the materials you will need to carry out your experiment.

Follow Your Plan

1. Make sure your teacher approves your plan before you start. Include any changes suggested by your teacher in your plans.
2. Carry out the experiment as planned.
3. **Record** any observations that you made while doing your experiment. Include suggestions for improving your design.

Analyze Your Data

1. **Compare** your car design to the designs of the other groups. What made the fastest car fast? What slowed the slowest car?
2. **Compare** your car's safety features to those of the other cars. What protected the eggs the best? How could you improve the unsuccessful designs?
3. **Predict** What effect would decreasing the speed of your car have on the safety of the egg?

Conclude and Apply

1. **Summarize** How did the best designs protect the egg?
2. **Apply** If you were designing cars, what could you do to better protect passengers from sudden stops?

Communicating Your Data

Write a descriptive paragraph about ways a car could be designed to protect its passengers effectively. Include a sketch of your ideas.

Oops! Accidents in SCIENCE

SOMETIMES GREAT DISCOVERIES HAPPEN BY ACCIDENT!

What Goes Around Comes Around
The Story of Boomerangs

Imagine a group gathered on a flat, yellow plain on the Australian Outback. One youth steps forward and, with the flick of an arm, sends a long, flat, angled stick soaring and spinning into the sky. The stick's path curves until it returns right back into the thrower's hand. Thrower after thrower steps forward, and the contest goes on all afternoon.

This contest involved throwing boomerangs—elegantly curved sticks. Because of how boomerangs are shaped, they always return to the thrower's hand

This amazing design is over 15,000 years old. Scientists believe that boomerangs developed from simple clubs thrown to stun and kill animals for food. Differently shaped clubs flew in different ways. As the shape of the club was refined, people probably started throwing them for fun too. In fact, today, using boomerangs for fun is still a popular sport, as world-class throwers compete in contests of strength and skill.

Boomerangs come in several forms, but all of them have several things in common. First a boomerang is shaped like an airplane's wing: flat on one side and curved on the other. Second, boomerangs are angled, which makes them spin as they fly. These two features determine the aerodynamics that give the boomerang its unique flight path.

From its beginning as a hunting tool to its use in today's World Boomerang Championships, the boomerang has remained a source of fascination for thousands of years.

Design Boomerangs are made from various materials. Research to find instructions for making boomerangs. After you and your friends build some boomerangs, have a competition of your own.

Science Online
For more information, visit
bookm.msscience.com/oops

Chapter 1 Study Guide

Reviewing Main Ideas

Section 1 — What is motion?

1. The position of an object depends on the reference point that is chosen.
2. An object is in motion if the position of the object is changing.
3. The speed of an object equals the distance traveled divided by the time:
$$s = \frac{d}{t}$$
4. The velocity of an object includes the speed and the direction of motion.
5. The motion of an object can be represented on a speed-time graph.

Section 2 — Acceleration

1. Acceleration is a measure of how quickly velocity changes. It includes a direction.
2. An object is accelerating when it speeds up, slows down, or turns.
3. When an object moves in a straight line, its acceleration can be calculated by
$$a = \frac{(s_f - s_i)}{t}$$

Section 3 — Momentum

1. Momentum equals the mass of an object times its velocity:
$$p = mv$$
2. Momentum is transferred from one object to another in a collision.
3. According to the law of conservation of momentum, the total amount of momentum of a group of objects doesn't change unless outside forces act on the objects.

Visualizing Main Ideas

Copy and complete the following table on motion.

Describing Motion

Quantity	Definition	Direction
Distance	length of path traveled	no
Displacement	direction and change in position	
Speed		no
Velocity	rate of change in position and direction	
Acceleration		
Momentum		yes

chapter 1 Review

Using Vocabulary

acceleration p. 14
average speed p. 11
inertia p. 19
instantaneous speed p. 11
law of conservation
   of momentum p. 21
mass p. 19
momentum p. 20
speed p. 10
velocity p. 13

Explain the relationship between each pair of terms.

1. speed—velocity
2. velocity—acceleration
3. velocity—momentum
4. momentum—law of conservation of momentum
5. mass—momentum
6. mass—inertia
7. momentum—inertia
8. average speed—instantaneous speed

Checking Concepts

Choose the word or phrase that best answers the question.

9. What measures the quantity of matter?
   A) speed
   B) weight
   C) acceleration
   D) mass

10. Which of the following objects is NOT accelerating?
    A) a jogger moving at a constant speed
    B) a car that is slowing down
    C) Earth orbiting the Sun
    D) a car that is speeding up

11. Which of the following equals speed?
    A) acceleration/time
    B) (change in velocity)/time
    C) distance/time
    D) displacement/time

12. Which of these is an acceleration?
    A) 5 m east
    B) 15 m/s east
    C) 52 m/s^2 east
    D) 32 s^2 east

13. Resistance to a change in motion increases when which of these increases?
    A) velocity
    B) speed
    C) instantaneous speed
    D) mass

14. What is 18 cm/h north an example of?
    A) speed
    B) velocity
    C) acceleration
    D) momentum

15. Which is true when the velocity and the acceleration of an object are in the same direction?
    A) The object's speed is constant.
    B) The object changes direction.
    C) The object speeds up.
    D) The object slows down.

16. Which of the following equals the change in velocity divided by the time?
    A) speed
    B) displacement
    C) momentum
    D) acceleration

17. You travel to a city 200 km away in 2.5 hours. What is your average speed in km/h?
    A) 180 km/h
    B) 12.5 km/h
    C) 80 km/h
    D) 500 km/h

18. A cue ball hits another billiard ball and slows down. Why does the speed of the cue ball decrease?
    A) The cue ball's momentum is positive.
    B) The cue ball's momentum is negative.
    C) Momentum is transferred to the cue ball.
    D) Momentum is transferred from the cue ball.

chapter 1 Review

Thinking Critically

19. **Explain** You run 100 m in 25 s. If you later run the same distance in less time, explain if your average speed increase or decrease.

Use the graph below to answer questions 20 and 21.

20. **Compare** For the motion of the object plotted on the speed-time graph above, how does the acceleration between 0 s and 3 s compare to the acceleration between 3 s and 5 s?

21. **Calculate** the acceleration of the object over the time interval from 0 s to 3 s.

22. **Infer** The molecules in a gas are often modeled as small balls. If the molecules all have the same mass, infer what happens if two molecules traveling at the same speed collide head on.

23. **Calculate** What is your displacement if you walk 100 m north, 20 m east, 30 m south, 50 m west, and then 70 m south?

24. **Infer** You are standing on ice skates and throw a basketball forward. Infer how your velocity after you throw the basketball compares with the velocity of the basketball.

25. **Determine** You throw a ball upward and then it falls back down. How does the velocity of the ball change as it rises and falls?

Use the graph below to answer question 26.

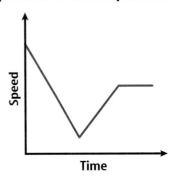

26. **Make and Use Graphs** The motion of a car is plotted on the speed-time graph above. Over which section of the graph is the acceleration of the car zero?

Performance Activities

27. **Demonstrate** Design a racetrack and make rules that specify the types of motion allowed. Demonstrate how to measure distance, measure time, and calculate speed accurately.

Applying Math

28. **Velocity of a Ball** Calculate the velocity of a 2-kg ball that has a momentum of 10 kg · m/s west.

29. **Distance Traveled** A car travels for a half hour at a speed of 40 km/h. How far does the car travel?

Use the graph below to answer question 30.

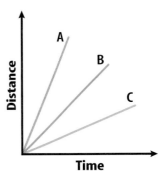

30. **Speed** From the graph determine which object is moving the fastest and which is moving the slowest.

Chapter 1 Standardized Test Practice

Part 1 Multiple Choice

Record your answers on the answer sheet provided by your teacher or on a sheet of paper.

1. What is the distance traveled divided by the time taken to travel that distance?
   A. acceleration   C. speed
   B. velocity   D. inertia

2. Sound travels at a speed of 330 m/s. How long does it take for the sound of thunder to travel 1,485 m?
   A. 45 s   C. 4,900 s
   B. 4.5 s   D. 0.22 s

Use the figure below to answer questions 3 and 4.

3. During which time period is the ball's average speed the fastest?
   A. between 0 and 1 s
   B. between 1 and 2 s
   C. between 2 and 3 s
   D. between 3 and 4 s

4. What is the average speed of the ball?
   A. 0.75 m/s   C. 10 m/s
   B. 1 m/s   D. 1.3 m/s

5. A car accelerates from 15 m/s to 30 m/s in 3.0 s. What is the car's acceleration?
   A. 10 m/s^2 C. 15 m/s^2
   B. 25 m/s^2 D. 5.0 m/s^2

6. Which of the following can occur when an object is accelerating?
   A. It speeds up.   C. It changes direction.
   B. It slows down.   D. all of the above

7. What is the momentum of a 21-kg bicycle traveling west at 3.0 m/s?
   A. 7 kg · m/s west   C. 18 kg · m/s west
   B. 63 kg · m/s west   D. 24 kg · m/s west

Use the figure below to answer questions 8–10.

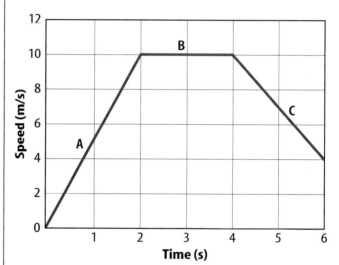

8. What is the acceleration between 0 and 2 s?
   A. 10 m/s^2 C. 0 m/s^2
   B. 5 m/s^2 D. −5 m/s^2

9. During what time period does the object have a constant speed?
   A. between 1 and 2 s
   B. between 2 and 3 s
   C. between 4 and 5 s
   D. between 5 and 6 s

10. What is the acceleration between 4 and 6 s?
    A. 10 m/s^2 C. 6 m/s^2
    B. 4 m/s^2 D. −3 m/s^2

11. An acorn falls from the top of an oak and accelerates at 9.8 m/s^2. It hits the ground in 1.5 s. What is the speed of the acorn when it hits the ground?
    A. 9.8 m/s   C. 15 m/s
    B. 20 m/s   D. 30 m/s

Standardized Test Practice

Part 2 — Short Response/Grid In

Record your answers on the answer sheet provided by your teacher or on a sheet of paper.

12. Do two objects that have the same mass always have the same momentum? Why or why not?

13. What is the momentum of a 57 kg cheetah running north at 27 m/s?

14. A sports car and a moving van are traveling at a speed of 30 km/h. Which vehicle will be easier to stop? Why?

Use the figure below to answer questions 15 and 16.

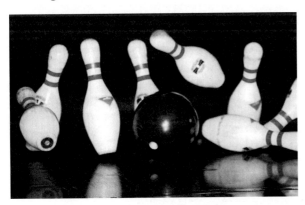

15. What happens to the momentum of the bowling ball when it hits the pins?

16. What happens to the speed of the ball and the speed of the pins?

17. What is the speed of a race horse that runs 1,500 m in 125 s?

18. A car travels for 5.5 h at an average speed of 75 km/h. How far did it travel?

19. If the speedometer on a car indicates a constant speed, can you be certain the car is not accelerating? Explain.

20. A girl walks 2 km north, then 2 km east, then 2 km south, then 2 km west. What distance does she travel? What is her displacement?

Part 3 — Open Ended

Record your answers on a sheet of paper.

Use the figure below to answer questions 21 and 22.

21. Describe the motion of the ball in terms of its speed, velocity, and acceleration.

22. During which part of its path does the ball have positive acceleration? During which part of its path does it have negative acceleration? Explain.

23. Describe what will happen when a baseball moving to the left strikes a bowling ball that is at rest.

24. A girl leaves school at 3:00 and starts walking home. Her house is 2 km from school. She gets home at 3:30. What was her average speed? Do you know her instantaneous speed at 3:15? Why or why not?

25. Why is it dangerous to try to cross a railroad track when a very slow-moving train is approaching?

Test-Taking Tip

Look for Missing Information Questions sometimes will ask about missing information. Notice what is missing as well as what is given.

chapter 2

Force and Newton's Laws

The BIG Idea
An object's motion changes if the forces acting on the object are unbalanced.

SECTION 1
Newton's First Law
Main Idea If the net force on an object is zero, the motion of the object does not change.

SECTION 2
Newton's Second Law
Main Idea An object's acceleration equals the net force divided by the mass.

SECTION 3
Newton's Third Law
Main Idea Forces act in equal but opposite pairs.

Moving at a Crawl
This enormous vehicle is a crawler that moves a space shuttle to the launch pad. The crawler and space shuttle together have a mass of about 7,700,000 kg. To move the crawler at a speed of about 1.5 km/h requires a force of about 10,000,000 N. This force is exerted by 16 electric motors in the crawler.

Science Journal Describe three examples of pushing or pulling an object. How did the object move?

Start-Up Activities

Forces and Motion

Imagine being on a bobsled team speeding down an icy run. Forces are exerted on the sled by the ice, the sled's brakes and steering mechanism, and gravity. Newton's laws predict how these forces cause the bobsled to turn, speed up, or slow down. Newton's Laws tell how forces cause the motion of any object to change.

1. Lean two metersticks parallel, less than a marble width apart on three books as shown on the left. This is your ramp.
2. Tap a marble so it rolls up the ramp. Measure how far up the ramp it travels before rolling back.
3. Repeat step 2 using two books, one book, and zero books. The same person should tap with the same force each time.
4. **Think Critically** Make a table to record the motion of the marble for each ramp height. What would happen if the ramp were perfectly smooth and level?

 Study Organizer

Newton's Laws Make the following Foldable to help you organize your thoughts about Newton's laws.

STEP 1 **Fold** a sheet of paper in half lengthwise. Make the back edge about 5 cm longer than the front edge.

STEP 2 **Turn** the paper so the fold is on the bottom. Then **fold** it into thirds.

STEP 3 **Unfold and cut** only the top layer along both folds to make three tabs.

STEP 4 **Label** the foldable as shown.

Make a Concept Map As you read the chapter, record what you learn about each of Newton's laws in your concept map.

 Preview this chapter's content and activities at bookm.msscience.com

Get Ready to Read

Compare and Contrast

① Learn It! Good readers compare and contrast information as they read. This means they look for similarities and differences to help them to remember important ideas. Look for signal words in the text to let you know when the author is comparing or contrasting.

Compare and Contrast Signal Words	
Compare	**Contrast**
as	but
like	or
likewise	unlike
similarly	however
at the same time	although
in a similar way	on the other hand

② Practice It! Read the excerpt below and notice how the author uses contrast signal words to describe the differences between weight and mass.

> When you stand on a bathroom scale, you are measuring the pull of the Earth's gravity — a force. **However,** mass is the amount of matter in an object, and doesn't depend on location. Weight will vary with location, **but** mass will remain constant.
>
> —*from page 44*

③ Apply It! Compare and contrast sliding friction on page 40 and air resistance on page 47.

Target Your Reading

Reading Tip

As you read, use other skills, such as summarizing and connecting, to help you understand comparisons and contrasts.

Use this to focus on the main ideas as you read the chapter.

① Before you read the chapter, respond to the statements below on your worksheet or on a numbered sheet of paper.
- Write an **A** if you **agree** with the statement.
- Write a **D** if you **disagree** with the statement.

② After you read the chapter, look back to this page to see if you've changed your mind about any of the statements.
- If any of your answers changed, explain why.
- Change any false statements into true statements.
- Use your revised statements as a study guide.

Science Online
Print out a worksheet of this page at bookm.msscience.com

Before You Read A or D		Statement	After You Read A or D
	1	If an object is moving, unbalanced forces are acting on the object.	
	2	When you jump up into the air, the ground exerts a force on you.	
	3	A force is a push or a pull.	
	4	Gravity does not pull on astronauts while in orbit around Earth.	
	5	Objects must be touching each other to apply forces on one another.	
	6	An object traveling in a circle at a constant speed is not accelerating.	
	7	Action and reaction force pairs cancel each other because they are equal in size but opposite in direction.	
	8	Gravity pulls on all objects that have mass.	
	9	An object at rest can have forces acting on it.	

section 1

Newton's First Law

as you read

What You'll Learn
- **Distinguish** between balanced and net forces.
- **Describe** Newton's first law of motion.
- **Explain** how friction affects motion.

Why It's Important
Forces can cause the motion of objects to change.

Review Vocabulary
velocity: the speed and direction of a moving object

New Vocabulary
- force
- net force
- balanced forces
- unbalanced forces
- Newton's first law of motion
- friction

Force

A soccer ball sits on the ground, motionless, until you kick it. Your science book sits on the table until you pick it up. If you hold your book above the ground, then let it go, gravity pulls it to the floor. In every one of these cases, the motion of the ball or book was changed by something pushing or pulling on it. An object will speed up, slow down, or turn only if something is pushing or pulling on it.

A **force** is a push or a pull. Examples of forces are shown in **Figure 1.** Think about throwing a ball. Your hand exerts a force on the ball, and the ball accelerates forward until it leaves your hand. After the ball leaves your hand, the force of gravity causes its path to curve downward. When the ball hits the ground, the ground exerts a force, stopping the ball.

A force can be exerted in different ways. For instance, a paper clip can be moved by the force a magnet exerts, the pull of Earth's gravity, or the force you exert when you pick it up. These are all examples of forces acting on the paper clip.

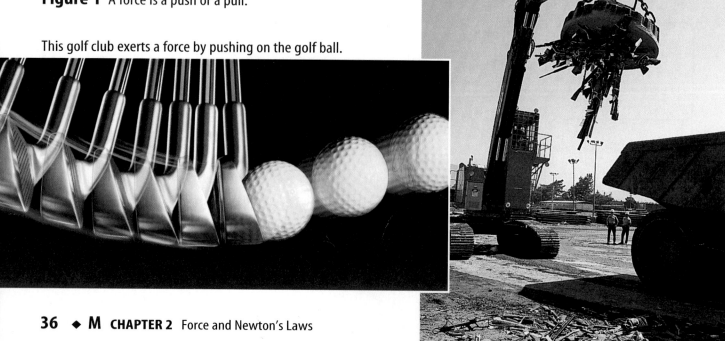

The magnet on the crane pulls the pieces of scrap metal upward.

Figure 1 A force is a push or a pull.

This golf club exerts a force by pushing on the golf ball.

This door is not moving because the forces exerted on it are equal and in opposite directions.

The door is closing because the force pushing the door closed is greater than the force pushing it open.

Figure 2 When the forces on an object are balanced, no change in motion occurs. A change in motion occurs only when the forces acting on an object are unbalanced.

Combining Forces More than one force can act on an object at the same time. If you hold a paper clip near a magnet, you, the magnet, and gravity all exert forces on the paper clip. The combination of all the forces acting on an object is the **net force.** When more than one force acts on an object, the net force determines how the motion of an object changes. If the motion of an object changes, its velocity changes. A change in velocity means the object is accelerating.

How do forces combine to form the net force? If the forces are in the same direction, they add together to form the net force. If two forces are in opposite directions, then the net force is the difference between the two forces, and it is in the direction of the larger force.

Balanced and Unbalanced Forces A force can act on an object without causing it to accelerate if other forces cancel the push or pull of the force. Look at **Figure 2.** If you and your friend push on a door with the same force in opposite directions, the door does not move. Because you both exert forces of the same size in opposite directions on the door, the two forces cancel each other. Two or more forces exerted on an object are **balanced forces** if their effects cancel each other and they do not change the object's velocity. If the forces on an object are balanced, the net force is zero. If the net force is not zero, the forces are **unbalanced forces.** Then the effects of the forces don't cancel, and the object's velocity changes.

Biomechanics Whether you run, jump, or sit, forces are being exerted on different parts of your body. Biomechanics is the study of how the body exerts forces and how it is affected by forces acting on it. Research how biomechanics has been used to reduce job-related injuries. Write a paragraph on what you've learned in your Science Journal.

Newton's First Law of Motion

If you stand on a skateboard and someone gives you a push, then you and your skateboard will start moving. You will begin to move when the force was applied. An object at rest—like you on your skateboard—remains at rest unless an unbalanced force acts on it and causes it to move.

Because a force had to be applied to make you move when you and your skateboard were at rest, you might think that a force has to be applied continually to keep an object moving. Surprisingly, this is not the case. An object can be moving even if the net force acting on it is zero.

The Italian scientist Galileo Galilei, who lived from 1564 to 1642, was one of the first to understand that a force doesn't need to be constantly applied to an object to keep it moving.

Galileo's ideas helped Isaac Newton to better understand the nature of motion. Newton, who lived from 1642 to 1727, explained the motion of objects in three rules called Newton's laws of motion.

Newton's first law of motion describes how an object moves when the net force acting on it is zero. According to **Newton's first law of motion,** if the net force acting on an object is zero, the object remains at rest, or if the object is already moving, continues to move in a straight line with constant speed.

Friction

Galileo realized the motion of an object doesn't change until an unbalanced force acts on it. Every day you see moving objects come to a stop. The force that brings nearly everything to a stop is **friction,** which is the force that acts to resist sliding between two touching surfaces, as shown in **Figure 3.** Friction is why you never see objects moving with constant velocity unless a net force is applied. Friction is the force that eventually brings your skateboard to a stop unless you keep pushing on it. Friction also acts on objects that are sliding or moving through substances such as air or water.

Figure 3 When two objects in contact try to slide past each other, friction keeps them from moving or slows them down.

Without friction, the rock climber would slide down the rock.

Friction slows down this sliding baseball player.

Friction Opposes Sliding Although several different forms of friction exist, they all have one thing in common. If two objects are in contact, frictional forces always try to prevent one object from sliding on the other object. If you rub your hand against a tabletop, you can feel the friction push against the motion of your hand. If you rub the other way, you can feel the direction of friction change so it is again acting against your hand's motion.

 What do the different forms of friction have in common?

Science Online

Topic: Galileo and Newton
Visit bookm.msscience.com for Web links to information about the lives of Galileo and Newton.

Activity Make a time line showing important events in the lives of either Galileo or Newton.

Older Ideas About Motion It took a long time to understand motion. One reason was that people did not understand the behavior of friction and that friction was a force. Because moving objects eventually come to a stop, people thought the natural state of an object was to be at rest. For an object to be in motion, something always had to be pushing or pulling it to keep the object moving. As soon as the force stopped, the object would stop moving.

Galileo understood that an object in constant motion is as natural as an object at rest. It was usually friction that made moving objects slow down and eventually come to a stop. To keep an object moving, a force had to be applied to overcome the effects of friction. If friction could be removed, an object in motion would continue to move in a straight line with constant speed. **Figure 4** shows motion where there is almost no friction.

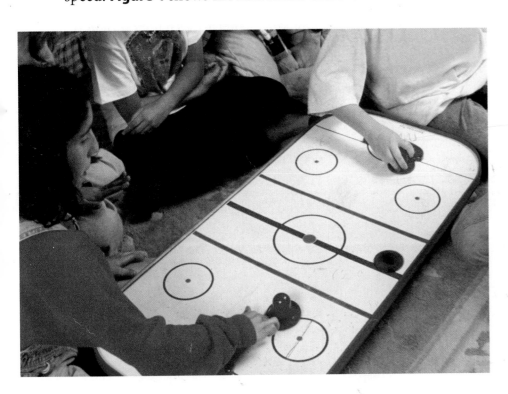

Figure 4 In an air hockey game, the puck floats on a layer of air, so that friction is almost eliminated. As a result, the puck moves in a straight line with nearly constant speed after it's been hit.
Infer *how the puck would move if there was no layer of air.*

SECTION 1 Newton's First Law

Observing Friction

Procedure

1. Lay a **bar of soap**, a **flat eraser**, and a **key** side by side on one end of a **hard-sided notebook**.
2. At a constant rate, slowly lift the end of notebook with objects on it. Note the order in which the objects start sliding.

Analysis

1. For which object was static friction the greatest? For which object was it the smallest? Explain, based on your observations.
2. Which object slid the fastest? Which slid the slowest? Explain why there is a difference in speed.
3. How could you increase and decrease the amount of friction between two materials?

Static Friction If you've ever tried pushing something heavy, like a refrigerator, you might have discovered that nothing happened at first. Then as you push harder and harder, the object suddenly will start to move. When you first start to push, friction between the heavy refrigerator and the floor opposes the force you are exerting and the net force is zero. The type of friction that prevents an object from moving when a force is applied is called static friction.

Static friction is caused by the attraction between the atoms on the two surfaces that are in contact. This causes the surfaces to stick or weld together where they are in contact. Usually, as the surface gets rougher and the object gets heavier, the force of static friction will be larger. To move the object, you have to exert a force large enough to break the bonds holding two surfaces together.

Sliding Friction While static friction acts on an object at rest, sliding friction slows down an object that slides. When you push a box over the foor, sliding friction acts in the direction opposite to the motion of the box. If you stop pushing, sliding friction causes the box to stop. Sliding friction is due to the microscopic roughness of two openers, as shown in **Figure 5.** The surfaces tend to stick together where they touch. The bonds between the surfaces are broken and form again as the surfaces slide past each other. This causes sliding friction. **Figure 6** shows that sliding friction is produced when the brake pad in a bicyle's brakes rub against the wheel.

Reading Check *What is the difference between static friction and sliding friction?*

Figure 5 Microscopic roughness, even on surfaces that seem smooth, such as the tray and metal shelf, causes sliding friction.

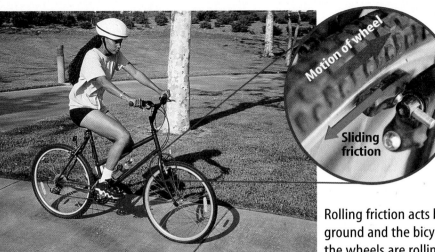

Figure 6 Sliding friction and rolling friction can act on a bicycle.

Sliding friction between the brake pads and the wheel causes the wheel to slow down.

Rolling friction acts between the ground and the bicycle tires when the wheels are rolling.

Rolling Friction If you're coasting on a bicycle or on a skateboard, you slow down and eventually stop because of another type of frictional force. Rolling friction occurs when an object rolls across a surface. It is rolling friction between the bicycle tires and the ground, as in **Figure 6,** that slows a moving bicycle.

The size of the rolling friction force due usually is much less than the force of sliding friction between the same surfaces. This is why it takes less force to pull a box on a wagon or cart with wheels, than to drag the box along the ground. Rolling friction between the wheels and the ground is less than the sliding friction between the box and the ground.

section 1 review

Summary

Force
- A force is a push or a pull.
- The net force on an object is the combination of all the forces acting on the object.
- The forces acting on an object can be balanced or unbalanced. If the forces are balanced, the net force is zero.

Newton's First Law of Motion
- If the net force on an object at rest is zero, the object remains at rest, or if the object is moving, it continues moving in a straight line with constant speed.

Friction
- Friction is the force that acts to resist sliding between two surfaces that are touching.
- Three types of friction are static friction, sliding friction, and rolling friction.

Self Check

1. **Explain** whether a force is acting on a car that is moving at 20 km/h and turns to the left.
2. **Describe** the factors that cause static friction between two surfaces to increase.
3. **Discuss** why friction made it difficult to discover Newton's first law of motion.
4. **Discuss** whether an object can be moving if the net force acting on the object is zero.
5. **Think Critically** For the following actions, explain whether the forces involved are balanced or unbalanced.
   a. You push a box until it moves.
   b. You push a box but it doesn't move.
   c. You stop pushing a box and it slows down.

Applying Skills

6. **Compare and contrast** static, sliding, and rolling friction.

section 2
Newton's Second Law

as you read

What You'll Learn
- **Explain** Newton's second law of motion.
- **Explain** why the direction of force is important.

Why It's Important
Newton's second law of motion explains how changes in motion can be calculated.

Review Vocabulary
acceleration: the change in velocity divided by the time over which the change occurred

New Vocabulary
- Newton's second law of motion
- weight
- center of mass

Force and Acceleration

When you go shopping in a grocery store and push a cart, you exert a force to make the cart move. If you want to slow down or change the direction of the cart, a force is required to do this, as well. Would it be easier for you to stop a full or empty grocery cart suddenly, as in **Figure 7**? When the motion of an object changes, the object is accelerating. Acceleration occurs any time an object speeds up, slows down, or changes its direction of motion.

Newton's second law of motion connects the net force on an object, its mass, and its acceleration. According to **Newton's second law of motion,** the acceleration of an object equals the net force divided by the mass and is in the direction of the net force. The acceleration can be calculated using this equation:

Newton's Second Law Equation

$$\text{acceleration (in meters/second}^2\text{)} = \frac{\text{net force (in newtons)}}{\text{mass (in kilograms)}}$$

$$a = \frac{F_{net}}{m}$$

In this equation, a is the acceleration, m is the mass, and F_{net} is the net force. If both sides of the above equation are multiplied by the mass, the equation can be written this way:

$$F_{net} = ma$$

Reading Check *What is Newton's second law?*

Figure 7 The force needed to change the motion of an object depends on its mass.
Predict *which grocery cart would be easier to stop.*

Units of Force Force is measured in newtons, abbreviated N. Because the SI unit for mass is the kilogram (kg) and acceleration has units of meters per second squared (m/s²), 1 N also is equal to 1 kg·m/s². In other words, to calculate a force in newtons from the equation shown on the prior page, the mass must be given in kg and the acceleration in m/s².

Gravity

One force that you are familiar with is gravity. Whether you're coasting down a hill on a bike or a skateboard or jumping into a pool, gravity is pulling you downward. Gravity also is the force that causes Earth to orbit the Sun and the Moon to orbit Earth.

What is gravity? The force of gravity exists between any two objects that have mass. Gravity always is attractive and pulls objects toward each other. A gravitational attraction exists between you and every object in the universe that has mass. However, the force of gravity depends on the mass of the objects and the distance between them. The gravitational force becomes weaker the farther apart the objects are and also decreases as the masses of the objects involved decrease.

For example, there is a gravitational force between you and the Sun and between you and Earth. The Sun is much more massive than Earth, but is so far away that the gravitational force between you and the Sun is too weak to notice. Only Earth is close enough and massive enough to exert a noticeable gravitational force on you. The force of gravity between you and Earth is about 1,650 times greater than between you and the Sun.

Newton and Gravity
Isaac Newton was the first to realize that gravity—the force that made objects fall to Earth—was also the force that caused the Moon to orbit Earth and the planets to orbit the Sun. In 1687, Newton published a book that included the law of universal gravitation. This law showed how to calculate the gravitational force between any two objects. Using the law of universal gravitation, astronomers were able to explain the motions of the planets in the solar system, as well as the motions of distant stars and galaxies.

Weight When you stand on a bathroom scale, what are you measuring? The **weight** of an object is the size of the gravitational force exerted on an object. Your weight on Earth is the gravitational force between you and Earth. On Earth, weight is calculated from this equation:

$$W = m\,(9.8 \text{ m/s}^2)$$

In this equation, W is the weight in N, and m is the mass in kg. Your weight would change if you were standing on a planet other than Earth, as shown in **Table 1.** Your weight on a different planet would be the gravitational force between you and the planet.

Table 1 Weight of 60-kg Person on Different Planets

Place	Weight in Newtons if Your Mass were 60 kg	Percent of Your Weight on Earth
Mars	221	37.7
Earth	588	100.0
Jupiter	1,390	236.4
Pluto	35	5.9

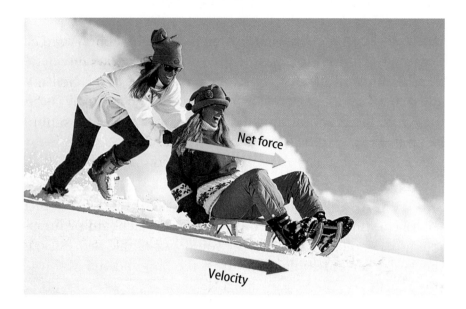

Figure 8 The sled speeds up when the net force on the sled is in the same direction as the sled's velocity.

Figure 9 The sled slows down when the net force on the sled is in the direction opposite to the sled's velocity.

Weight and Mass Weight and mass are different. Weight is a force, just like the push of your hand is a force, and is measured in newtons. When you stand on a bathroom scale, you are measuring the pull of Earth's gravity—a force. However, mass is the amount of matter in an object, and doesn't depend on location. Weight will vary with location, but mass will remain constant. A book with a mass of 1 kg has a mass of 1 kg on Earth or on Mars. However, the weight of the book would be different on Earth and Mars. The two planets would exert a different gravitational force on the book.

Using Newton's Second Law

The second law of motion tells how to calculate the acceleration of an object if the object's mass and the forces acting on it are known. Recall that the acceleration equals the change in velocity divided by the change in time. If an object's acceleration is known, the change in its velocity can be determined

Speeding Up When does an unbalanced force cause an object to speed up? If an object is moving, a net force applied in the same direction as the object is moving causes the object to speed up. For exmple, in **Figure 8** the applied force is in the same direction as the sled's velocity. This makes the sled speed up and its velocity increase.

The net force on a ball falling to the ground is downward. This force is in the same direction that the ball is moving. Because the net force on the ball is in the same firecion as the ball's velocity, the ball speeds up as it falls.

Slowing Down If the net force on an object is in the direction opposite to the object's velocity, the object slows down. In **Figure 9,** the force of sliding friction becomes larger when the boy puts his feet in the snow. The net force on the sled is the cmobination of gravity and sliding friction. When the sliding friction force becomes lage enough, the net force is opposite to the sled's velocity. This causes the sled to slow down.

Calculating Acceleration Newton's second law of motion can be used to calculate acceleration. For example, suppose you pull a 10-kg box so that the net force on the box is 5 N. The acceleration can be found as follows:

$$a = \frac{F_{net}}{m} = \frac{5 \text{ N}}{10 \text{ kg}} = 0.5 \text{ m/s}^2$$

The box keeps accelerating as long as you keep pulling on it. The acceleration does not depend on how fast the box is moving. It depends only on the net force and the mass of the box.

Applying Math — Solve a Simple Equation

ACCELERATION OF A CAR A net force of 4,500 N acts on a car with a mass of 1,500 kg. What is the acceleration of the car?

Solution

1 *This is what you know:*
- net force: $F_{net} = 4,500$ N
- mass: $m = 1,500$ kg

2 *This is what you need to find:* acceleration: $a = ?$ m/s^2

3 *This is the procedure you need to use:* Substitute the known values for net force and mass into the equation for Newton's second law of motion to calculate the acceleration:

$$a = \frac{F_{net}}{m} = \frac{4,500 \text{ N}}{1,500 \text{ kg}} = 3.0 \frac{\text{N}}{\text{kg}} = 3.0 \text{ m/s}^2$$

4 *Check your answer:* Multiply your answer by the mass, 1,500 kg. The result should be the given net force, 4,500 N.

Practice Problems

1. A book with a mass of 2.0 kg is pushed along a table. If the net force on the book is 1.0 N, what is the book's acceleration?

2. A baseball has a mass of 0.15 kg. What is the net force on the ball if its acceleration is 40.0 m/s^2?

For more practice visit
bookm.msscience.com/
math_practice

Figure 10 The force due to gravity on the ball is at an angle to the ball's velocity. This causes the ball to move in a curved path. **Infer** *how the ball would move if it were thrown horizontally.*

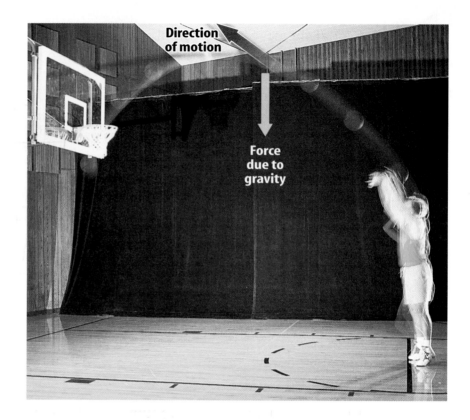

Turning Sometimes te net force on an object and the object's velocity, are neither in the same direction nor in the opposite direction. Then the object will move in a curved path, instead of a straight line.

When you shoot a basketball, it doesn't continue to move in a straight line after it leaves your hand. Instead the ball starts to curve downward, as shown in **Figure 10.** Gravity pulls the ball downward,.so the net force on the ball is at an angle to the ball's velocity. This causes the ball to move in a curved path.

Circular Motion

A rider on a merry-go-round ride moves in a circle. This type of motion is called circular motion. If you are in circular motion, your direction of motion is constantly changing. This means you are constantly accelerating. According to Newton's second law of motion, if you are constantly accelerating, there must be a non-zero net force acting on you the entire time.

To cause an object to move in circular motion with constant speed, the net force on the object must be at right angles to the velocity. When an object moves in circular motion, the net force on the object is called centripetal force. The direction of the centripetal force is toward the center of the object's circular path.

46 ◆ **M** CHAPTER 2 Force and Newton's Laws

Satellite Motion Objects that orbit Earth are satellites of Earth. Some satellites go around Earth in nearly circular orbits. The centripetal force is due to gravity between the satellite and Earth. Gravity causes the net force on the satellite to always point toward Earth, which is center of the satellite's orbit. Why doesn't a satellite fall to Earth like a baseball does? Actually, a satellite is falling to Earth just like a baseball.

Suppose Earth were perfectly smooth and you throw a baseball horizontally. Gravity pulls the baseball downward so it travels in a curved path. If the baseball is thrown faster, its path is less curved, and it travels farther before it hits the ground. If the baseball were traveling fast enough, as it fell, its curved path would follow the curve of Earth's surface as shown in **Figure 11.** Then the baseball would never hit the ground. Instead, it would continue to fall around Earth.

Satellites in orbit are being pulled toward Earth just as baseballs are. The difference is that satellites are moving so fast horizontally that Earth's surface curves downward at the same rate that the satellites are falling downward. The speed at which a object must move to go into orbit near Earth's surface is about 8 km/s, or about 29,000 km/h.

Figure 11 The faster a ball is thrown, the farther it travels before gravity pulls it to Earth. If the ball is traveling fast enough, Earth's surface curves away from it as fast as it falls downward. Then the ball never hits the ground.

Air Resistance

Whether you are walking, running, or biking, air is pushing against you. This push is air resistance. Air resistance is a form of friction that acts to slow down any object moving in the air. Air resistance is a force that gets larger as an object moves faster. Air resistance also depends on the shape of an object. A piece of paper crumpled into a ball falls faster than a flat piece of paper falls.

When an object falls it speeds up as gravity pulls it downward. At the same time, the force of air resistance pushing up on the object is increasing as the object moves faster. Finally, the upward air resistance force becomes large enough to equal the downward force of gravity.

When the air resistance force equals the weight, the net force on the object is zero. By Newton's second law, the object's acceleration then is zero, and its speed no longer increases. When air resistance balances the force of gravity, the object falls at a constant speed called the terminal velocity.

Figure 12 The wrench is spinning as it slides across the table. The center of mass of the wrench, shown by the dots, moves as if the net force is acting at that point.

Center of Mass

When you throw a stick, the motion of the stick might seem to be complicated. However, there is one point on the stick, called the center of mass, that moves in a smooth path. The **center of mass** is the point in an object that moves as if all the object's mass were concentrated at that point. For a symmetrical object, such as a ball, the center of mass is at the object's center. However, for any object the center of mass moves as if the net force is being applied there.

Figure 12 shows how the center of mass of a wrench moves as it slides across a table. The net force on the wrench is the force of friction between the wrench and the table. This causes the center of mass to move in a straight line with decreasing speed.

section 2 review

Summary

Force and Acceleration
- According to Newton's second law, the net force on an object, its mass, and its acceleration are related by
  $$F_{net} = ma$$

Gravity
- The force of gravity between any two objects is always attractive and depends on the masses of the objects and the distance between them.

Using Newton's Second Law
- A moving object speeds up if the net force is in the direction of the motion.
- A moving object slows down if the net force is in the direction opposite to the motion.
- A moving object turns if the net force is at an angle to the direction of motion.

Circular Motion
- In circular motion with constant speed, the net force is called the centripetal force and points toward the center of the circular path.

Self Check

1. **Make a diagram** showing the forces acting on a coasting bike rider traveling at 25 km/h on a flat roadway.
2. **Analyze** how your weight would change with time if you were on a space ship traveling away from Earth toward the Moon.
3. **Explain** how the force of air resistance depends on an object's speed.
4. **Infer** the direction of the net force acting on a car as it slows down and turns right.
5. **Think Critically** Three students are pushing on a box. Under what conditions will the motion of the box change?

Applying Math

6. **Calculate Net Force** A car has a mass of 1,500 kg. If the car has an acceleration of 2.0 m/s², what is the net force acting on the car?
7. **Calculate Mass** During a softball game, a softball is struck by a bat and has an acceleration of 1,500 m/s². If the net force exerted on the softball by the bat is 300 N, what is the softball's mass?

bookm.msscience.com/self_check_quiz

section 3
Newton's Third Law

Action and Reaction

Newton's first two laws of motion explain how the motion of a single object changes. If the forces acting on the object are balanced, the object will remain at rest or stay in motion with constant velocity. If the forces are unbalanced, the object will accelerate in the direction of the net force. Newton's second law tells how to calculate the acceleration, or change in motion, of an object if the net force acting on it is known.

Newton's third law describes something else that happens when one object exerts a force on another object. Suppose you push on a wall. It may surprise you to learn that if you push on a wall, the wall also pushes on you. According to **Newton's third law of motion**, forces always act in equal but opposite pairs. Another way of saying this is for every action, there is an equal but opposite reaction. This means that when you push on a wall, the wall pushes back on you with a force equal in strength to the force you exerted. When one object exerts a force on another object, the second object exerts the same size force on the first object, as shown in **Figure 13**.

What You'll Learn
- Identify the relationship between the forces that objects exert on each other.

Why It's Important
Newton's third law can explain how birds fly and rockets move.

Review Vocabulary
force: a push or a pull

New Vocabulary
- Newton's third law of motion

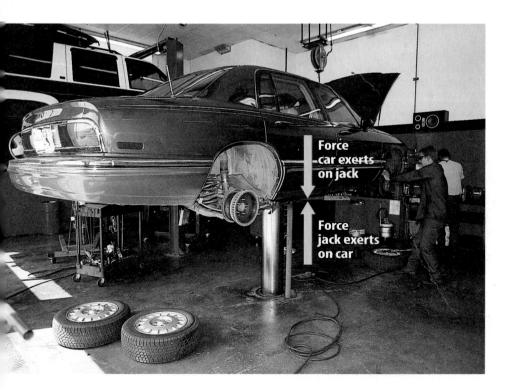

Figure 13 The car jack is pushing up on the car with the same amount of force with which the car is pushing down on the jack. **Identify** *the other force acting on the car.*

Figure 14 In this collision, the first car exerts a force on the second. The second exerts the same force in the opposite direction on the first car.
Explain whether both cars will have the same acceleration.

Topic: How Birds Fly
Visit bookm.msscience.com for Web links to information about how birds and other animals fly.

Activity Make a diagram showing the forces acting on a bird as it flies.

Figure 15 When the child pushes against the wall, the wall pushes against the child.

Action and Reaction Forces Don't Cancel The forces exerted by two objects on each other are often called an action-reaction force pair. Either force can be considered the action force or the reaction force. You might think that because action-reaction forces are equal and opposite that they cancel. However, action and reaction force pairs don't cancel because they act on different objects. Forces can cancel only if they act on the same object.

For example, imagine you're driving a bumper car and are about to bump a friend in another car, as shown in **Figure 14.** When the two cars collide, your car pushes on the other car. By Newton's third law, that car pushes on your car with the same force, but in the opposite direction. This force causes you to slow down. One force of the action-reaction force pair is exerted on your friend's car, and the other force of the force pair is exerted on your car. Another example of an action-reaction pair is shown in **Figure 15.**

You constantly use action-reaction force pairs as you move about. When you jump, you push down on the ground. The ground then pushes up on you. It is this upward force that pushes you into the air. **Figure 16** shows some examples of how Newton's laws of motion are demonstrated in sporting events.

 Birds and other flying creatures also use Newton's third law. When a bird flies, its wings push in a downward and a backward direction. This pushes air downward and backward. By Newton's third law, the air pushes back on the bird in the opposite directions—upward and forward. This force keeps a bird in the air and propels it forward.

NATIONAL GEOGRAPHIC VISUALIZING NEWTON'S LAWS IN SPORTS

Figure 16

Although it is not obvious, Newton's laws of motion are demonstrated in sports activities all the time. According to the first law, if an object is in motion, it moves in a straight line with constant speed unless a net force acts on it. If an object is at rest, it stays at rest unless a net force acts on it. The second law states that a net force acting on an object causes the object to accelerate in the direction of the force. The third law can be understood this way—for every action force, there is an equal and opposite reaction force.

▶ **NEWTON'S SECOND LAW** As Tiger Woods hits a golf ball, he applies a force that will drive the ball in the direction of that force—an example of Newton's second law.

▲ **NEWTON'S FIRST LAW** According to Newton's first law, the diver does not move in a straight line with constant speed because of the force of gravity.

▶ **NEWTON'S THIRD LAW** Newton's third law applies even when objects do not move. Here a gymnast pushes downward on the bars. The bars push back on the gymnast with an equal force.

Figure 17 The force of the ground on your foot is equal and opposite to the force of your foot on the ground. If you push back harder, the ground pushes forward harder.
Determine *In what direction does the ground push on you if you are standing still?*

Change in Motion Depends on Mass Often you might not notice the effects of the forces in an action-reaction pair. This can hapen if one of the objects involved is much more massive than the other. Then the massice object might seem to remain motionless when one of the action-reaction forces acts on it. For example, when you walk forward, as in **Figure 17,** you push backward on the ground. Earth then pushes you forward with the same size force. Because Earth's mass is so large, the force you exert causes Earth to have an extremely small acceleration. This acceleration is so small that Earth's change in motion is undetectable as you walk.

A Rocket Launch The launching of a space shuttle is a spectacular example of Newton's third law. Three rocket engines supply the force, called thrust, that lifts the rocket. When the rocket fuel is ignited, a hot gas is produced. The gas molecules collide with the inside of the engine, as shown in **Figure 18.** As these collisions occur, the engine pushes the hot gases downward. According to Newton's third law of motion, the hot gases push upward on the engine. The upward force exerted by the gases on the rocket propels the rocket upward.

Figure 18 Newton's third law enables a rocket to fly. The rocket pushes the gas molecules downward, and the gas molecules push the rocket upward.

Figure 19 Whether you are standing on Earth or falling, the force of Earth's gravity on you doesn't change. However, your weight measured by a scale would change.

Weightlessness

You might have seen pictures of astronauts floating inside a space shuttle as it orbits Earth. The astronauts are said to be weightless, as if Earth's gravity were no longer pulling on them. Yet the force of gravity on the shuttle is almost 90 percent as large as at Earth's surface. Newton's laws of motion can explain why the astronauts float as if there were no forces acting on them.

Measuring Weight Think about how you measure your weight. When you stand on a scale, your weight pushes down on the scale. This causes the scale pointer to point to your weight. At the same time, by Newton's third law the scale pushes up on you with a force equal to your weight, as shown in **Figure 19**. This force balances the downward pull of gravity on you.

Free Fall and Weightlessness Now suppose you were standing on a scale in an elevator that is falling, as shown in **Figure 19**. A falling object is in free fall when the only force acting on the object is gravity. Inside the free-falling elevator, you and the scale are both in free fall. Because the only force acting on you is gravity, the scale no longer is pushing up on you. According to Newton's third law, you no longer push down on the scale. So the scale pointer stays at zero and you seem to be weightless. Weightlessness is the condition that occurs in free fall when the weight of an object seems to be zero.

However, you are not really weightless in free fall because Earth is still pulling down on you. With nothing to push up on you, such as your chair, you would have no sensation of weight.

Measuring Force Pairs

Procedure
1. Work in pairs. Each person needs a **spring scale**.
2. Hook the two scales together. Each person should pull back on a scale. Record the two readings. Pull harder and record the two readings.
3. Continue to pull on both scales, but let the scales move toward one person. Do the readings change?
4. Try to pull in such a way that the two scales have different readings.

Analysis
1. What can you conclude about the pair of forces in each situation?
2. Explain how this experiment demonstrates Newton's third law.

Weightlessness in Orbit To understand how objects move in the orbiting space shuttle, imagine you were holding a ball in the free-falling elevator. If you let the ball go, the position of the ball relative to you and the elevator wouldn't change, because you, the ball, and the elevator are moving at the same speed.

However, suppose you give the ball a gentle push downward. While you are pushing the ball, this downward force adds to the downward force of gravity. According to Newton's second law, the acceleration of the ball increases. So while you are pushing, the acceleration of the ball is greater than the acceleration of both you and the elevator. This causes the ball to speed up relative to you and the elevator. After it speeds up, it continues moving faster than you and the elevator, and it drifts downward until it hits the elevator floor.

When the space shuttle orbits Earth, the shuttle and all the objects in it are in free fall. They are falling in a curved path around Earth, instead of falling straight downward. As a result, objects in the shuttle appear to be weightless, as shown in **Figure 20.** A small push causes an object to drift away, just as a small downward push on the ball in the free-falling elevator caused it to drift to the floor.

Figure 20 These oranges seem to be floating because they are falling around Earth at the same speed as the space shuttle and the astronauts. As a result, they aren't moving relative to the astronauts in the cabin.

section 3 review

Summary

Action and Reaction

- According to Newton's third law, when one object exerts a force on another object, the second object exerts the same size force on the first object.
- Either force in an action-reaction force pair can be the action force or the reaction force.
- Action and reaction force pairs don't cancel because they are exerted on different objects.
- When action and reaction forces are exerted by two objects, the accelerations of the objects depend on the masses of the objects.

Weightlessness

- A falling object is in free fall if the only force acting on it is gravity.
- Weightlessness occurs in free fall when the weight of an object seems to be zero.
- Objects orbiting Earth appear to be weightless because they are in free fall in a curved path around Earth.

Self Check

1. **Evaluate** the force a skateboard exerts on you if your mass is 60 kg and you push on the skateboard with a force of 60 N.
2. **Explain** why you move forward and a boat moves backward when you jump from a boat to a pier.
3. **Describe** the action and reaction forces when a hammer hits a nail.
4. **Infer** You and a child are on skates and you give each other a push. If the mass of the child is half your mass, who has the greater acceleration? By what factor?
5. **Think Critically** Suppose you are walking in an airliner in flight. Use Newton's third law to describe the effect of your walk on the motion on the airliner.

Applying Math

6. **Calculate Acceleration** A person standing in a canoe exerts a force of 700 N to throw an anchor over the side. Find the acceleration of the canoe if the total mass of the canoe and the person is 100 kg.

BALLOON RACES

Real-World Question

The motion of a rocket lifting off a launch pad is determined by Newton's laws of motion. Here you will make a balloon rocket that is powered by escaping air. How do Newton's laws of motion explain the motion of balloon rockets?

Goals
- **Measure** the speed of a balloon rocket.
- **Describe** how Newton's laws explain a rocket's motion.

Materials
balloons   meterstick
drinking straws   stopwatch
string   *clock
tape   *Alternate materials

Safety Precautions

Procedure

1. Make a rocket path by threading a string through a drinking straw. Run the string across the classroom and fasten at both ends.
2. Blow up a balloon and hold it tightly at the end to prevent air from escaping. Tape the balloon to the straw on the string.
3. Release the balloon so it moves along the string. Measure the distance the balloon travels and the time it takes.
4. Repeat steps 2 and 3 with different balloons.

Conclude and Apply

1. **Compare and contrast** the distances traveled. Which rocket went the greatest distance?
2. **Calculate** the average speed for each rocket. Compare and contrast them. Which rocket has the greatest average speed?
3. **Infer** which aspects of these rockets made them travel far or fast.
4. **Draw** a diagram showing all the forces acting on a balloon rocket.
5. Use Newton's laws of motion to explain the motion of a balloon rocket from launch until it comes to a stop.

Communicating Your Data

Discuss with classmates which balloon rocket traveled the farthest. Why? **For more help, refer to the** Science Skill Handbook.

LAB **M** ◆ 55

LAB Design Your Own

MODELING MOTION IN TWO DIRECTIONS

Goals
- **Move** the skid across the ground using two forces.
- **Measure** how fast the skid can be moved.
- **Determine** how smoothly the direction can be changed.

Possible Materials
masking tape
stopwatch
* watch or clock with a second hand
meterstick
*metric tape measure
spring scales marked in newtons (2)
plastic lid
golf ball
*tennis ball
*Alternate materials

Safety Precautions

Real-World Question

When you move a computer mouse across a mouse pad, how does the rolling ball tell the computer cursor to move in the direction that you push the mouse? Inside the housing for the mouse's ball are two or more rollers that the ball rubs against as you move the mouse. They measure up-and-down and back-and-forth motions. The motion of the cursor on the screen is based on the movement of the up-and-down rollers and the back-and-forth rollers. Can any object be moved along a path by a series of motions in only two directions?

Form a Hypothesis

How can you combine forces to move in a straight line, along a diagonal, or around corners? Place a golf ball on something that will slide, such as a plastic lid. The plastic lid is called a skid. Lay out a course to follow on the floor. Write a plan for moving your golf ball along the path without having the golf ball roll away.

Test Your Hypothesis

Make a Plan

1. Lay out a course that involves two directions, such as always moving forward or left.
2. Attach two spring scales to the skid. One always will pull straight forward. One always will pull to one side. You cannot turn the skid. If one scale is pulling toward the door of your classroom, it always must pull in that direction. (It can pull with zero force if needed, but it can't push.)
3. How will you handle movements along diagonals and turns?
4. How will you measure speed?

56 ◆ **M CHAPTER 2** Force and Newton's Laws

Using Scientific Methods

5. **Experiment** with your skid. How hard do you have to pull to counteract sliding friction at a given speed? How fast can you accelerate? Can you stop suddenly without spilling the golf ball, or do you need to slow down?

6. **Write** a plan for moving your golf ball along the course by pulling only forward or to one side. Be sure you understand your plan and have considered all the details.

Follow Your Plan

1. Make sure your teacher approves your plan before you start.
2. Move your golf ball along the path.
3. Modify your plan, if needed.
4. **Organize** your data so they can be used to run your course and write them in your Science Journal.
5. **Test** your results with a new route.

Analyze Your Data

1. What was the difference between the two routes? How did this affect the forces you needed to use on the golf ball?
2. How did you separate and control variables in this experiment?
3. Was your hypothesis supported? Explain.

Conclude and Apply

1. What happens when you combine two forces at right angles?
2. If you could pull on all four sides (front, back, left, right) of your skid, could you move anywhere along the floor? Make a hypothesis to explain your answer.

Compare your conclusions with those of other students in your class. **For more help, refer to the** Science Skill Handbook.

LAB **M** ◆ 57

TIME SCIENCE AND Society
SCIENCE ISSUES THAT AFFECT YOU!

Air Bag Safety

After complaints and injuries, air bags in cars are helping all passengers

The car in front of yours stops suddenly. You hear the crunch of car against car and feel your seat belt grab you. Your mom is covered with, not blood, thank goodness, but with a big white cloth. Your seat belts and air bags worked perfectly.

Popcorn in the Dash

Air bags have saved more than a thousand lives since 1992. They are like having a giant popcorn kernel in the dashboard that pops and becomes many times its original size. But unlike popcorn, an air bag is triggered by impact, not temperature. In a crash, a chemical reaction produces a gas that expands in a split second, inflating a balloonlike bag to cushion the driver and possibly the front-seat passenger. The bag deflates quickly so it doesn't trap people in the car.

Newton and the Air Bag

When you're traveling in a car, you move with it at whatever speed it is going. According to Newton's first law, you are the object in motion, and you will continue in motion unless acted upon by a force, such as a car crash.

Unfortunately, a crash stops the car, but it doesn't stop you, at least, not right away. You continue moving forward if your car doesn't have air bags or if you haven't buckled your seat belt. You stop when you strike the inside of the car. You hit the dashboard or steering wheel while traveling at the speed of the car. When an air bag inflates, you come to a stop more slowly, which reduces the force that is exerted on you.

A test measures the speed at which an air bag deploys.

Measure Hold a paper plate 26 cm in front of you. Use a ruler to measure the distance. That's the distance drivers should have between the chest and the steering wheel to make air bags safe. Inform adult drivers in your family about this safety distance.

For more information, visit
bookm.msscience.com/time

Chapter 2 Study Guide

Reviewing Main Ideas

Section 1 — Newton's First Law

1. A force is a push or a pull.
2. Newton's first law states that objects in motion tend to stay in motion and objects at rest tend to stay at rest unless acted upon by a nonzero net force.
3. Friction is a force that resists motion between surfaces that are touching each other.

Section 2 — Newton's Second Law

1. Newton's second law states that an object acted upon by a net force will accelerate in the direction of this force.
2. The acceleration due to a net force is given by the equation $a = F_{net}/m$.

3. The force of gravity between two objects depends on their masses and the distance between them.
4. In circular motion, a force pointing toward the center of the circle acts on an object.

Section 3 — Newton's Third Law

1. According to Newton's third law, the forces two objects exert on each other are always equal but in opposite directions.
2. Action and reaction forces don't cancel because they act on different objects.
3. Objects in orbit appear to be weightless because they are in free fall around Earth.

Visualizing Main Ideas

Copy and complete the following concept map on Newton's laws of motion.

chapter 2 Review

Using Vocabulary

balanced forces p. 37
center of mass p. 48
force p. 36
friction p. 38
net force p. 37
Newton's first law of motion p. 38
Newton's second law of motion p. 42
Newton's third law of motion p. 49
unbalanced forces p. 37
weight p. 43

Explain the differences between the terms in the following sets.

1. force—inertia—weight
2. Newton's first law of motion—Newton's third law of motion
3. friction—force
4. net force—balanced forces
5. weight—weightlessness
6. balanced forces—unbalanced forces
7. friction—weight
8. Newton's first law of motion—Newton's second law of motion
9. friction—unbalanced force
10. net force—Newton's third law of motion

Checking Concepts

Choose the word or phrase that best answers the question.

11. Which of the following changes when an unbalanced force acts on an object?
    A) mass   C) inertia
    B) motion   D) weight

12. Which of the following is the force that slows a book sliding on a table?
    A) gravity
    B) static friction
    C) sliding friction
    D) inertia

Use the illustration below to answer question 13.

13. Two students are pushing on the left side of a box and one student is pushing on the right. The diagram above shows the forces they exert. Which way will the box move?
    A) up   C) down
    B) left   D) right

14. What combination of units is equivalent to the newton?
    A) m/s^2 C) $kg·m/s^2$
    B) $kg·m/s$ D) kg/m

15. Which of the following is a push or a pull?
    A) force   C) acceleration
    B) momentum   D) inertia

16. An object is accelerated by a net force in which direction?
    A) at an angle to the force
    B) in the direction of the force
    C) in the direction opposite to the force
    D) Any of these is possible.

17. You are riding on a bike. In which of the following situations are the forces acting on the bike balanced?
    A) You pedal to speed up.
    B) You turn at constant speed.
    C) You coast to slow down.
    D) You pedal at constant speed.

18. Which of the following has no direction?
    A) force   C) weight
    B) acceleration D) mass

chapter 2 Review

Thinking Critically

19. **Explain** why the speed of a sled increases as it moves down a snow-covered hill, even though no one is pushing on the sled.

20. **Explain** A baseball is pitched east at a speed of 40 km/h. The batter hits it west at a speed of 40 km/h. Did the ball accelerate?

21. **Form a Hypothesis** Frequently, the pair of forces acting between two objects are not noticed because one of the objects is Earth. Explain why the force acting on Earth isn't noticed.

22. **Identify** A car is parked on a hill. The driver starts the car, accelerates until the car is driving at constant speed, drives at constant speed, and then brakes to put the brake pads in contact with the spinning wheels. Explain how static friction, sliding friction, rolling friction, and air resistance are acting on the car.

23. **Draw Conclusions** You hit a hockey puck and it slides across the ice at nearly a constant speed. Is a force keeping it in motion? Explain.

24. **Infer** Newton's third law describes the forces between two colliding objects. Use this connection to explain the forces acting when you kick a soccer ball.

25. **Recognize Cause and Effect** Use Newton's third law to explain how a rocket accelerates upon takeoff.

26. **Predict** Two balls of the same size and shape are dropped from a helicopter. One ball has twice the mass of the other ball. On which ball will the force of air resistance be greater when terminal velocity is reached?

Use the figure below to answer question 27.

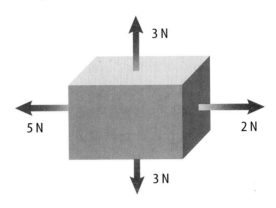

27. **Interpreting Scientific Illustrations** Is the force on the box balanced? Explain.

Performance Activities

28. **Oral Presentation** Research one of Newton's laws of motion and compose an oral presentation. Provide examples of the law. You might want to use a visual aid.

29. **Writing in Science** Create an experiment that deals with Newton's laws of motion. Document it using the following subject heads: *Title of Experiment, Partners' Names, Hypothesis, Materials, Procedures, Data, Results,* and *Conclusion*.

Applying Math

30. **Acceleration** If you exert a net force of 8 N on a 2-kg object, what will its acceleration be?

31. **Force** You push against a wall with a force of 5 N. What is the force the wall exerts on your hands?

32. **Net Force** A 0.4-kg object accelerates at 2 m/s^2. Find the net force.

33. **Friction** A 2-kg book is pushed along a table with a force of 4 N. Find the frictional force on the book if the book's acceleration is 1.5 m/s^2.

chapter 2 Standardized Test Practice

Part 1 Multiple Choice

Record your answers on the answer sheet provided by your teacher or on a sheet of paper.

1. Which of the following descriptions of gravitational force is *not* true?
   A. It depends on the mass of objects.
   B. It is a repulsive force.
   C. It depends on the distance between objects.
   D. It exists between all objects.

Use the table below to answer questions 2 and 3.

Mass of Common Objects	
Object	Mass (g)
Cup	380
Book	1,100
Can	240
Ruler	25
Stapler	620

2. Which object would have an acceleration of 0.89 m/s² if you pushed on it with a force of 0.55 N?
   A. book   C. ruler
   B. can   D. stapler

3. Which object would have the greatest acceleration if you pushed on it with a force of 8.2 N?
   A. can   C. ruler
   B. stapler   D. book

Test-Taking Tip

Check Symbols Be sure you understand all symbols on a table or graph before answering any questions about the table or graph.

Question 3 The mass of the objects are given in grams, but the force is given in newtons which is a kg·m/s². The mass must be converted from grams to kilograms.

4. What is the weight of a book that has a mass of 0.35 kg?
   A. 0.036 N   C. 28 N
   B. 3.4 N   D. 34 N

5. If you swing an object on the end of a string around in a circle, the string pulls on the object to keep it moving in a circle. What is the name of this force?
   A. inertial   C. resistance
   B. centripetal   D. gravitational

6. What is the acceleration of a 1.4-kg object if the gravitational force pulls downward on it, but air resistance pushes upward on it with a force of 2.5 N?
   A. 11.6 m/s², downward
   B. 11.6 m/s², upward
   C. 8.0 m/s², downward
   D. 8.0 m/s², upward

Use the figure below to answer questions 7 and 8.

7. The figure above shows the horizontal forces that act on a box that is pushed from the left with a force of 12 N. What force is resisting the horizontal motion in this illustration?
   A. friction   C. inertia
   B. gravity   D. momentum

8. What is the acceleration of the box?
   A. 27 m/s²   C. 4.3 m/s²
   B. 4.8 m/s²   D. 0.48 m/s²

Standardized Test Practice

Part 2 Short Response/Grid In

Record your answers on the answer sheet provided by your teacher or on a sheet of paper.

9. A skater is coasting along the ice without exerting any apparent force. Which law of motion explains the skater's ability to continue moving?

10. After a soccer ball is kicked into the air, what force or forces are acting on it?

11. What is the force on an 8.55-kg object that accelerates at 5.34 m/s^2.

Use the figure below to answer questions 12 and 13.

12. Two bumper cars collide and then move away from each other. How do the forces the bumper cars exert on each other compare?

13. After the collision, determine whether both bumper cars will have the same acceleration.

14. Does acceleration depend on the speed of an object? Explain.

15. An object acted on by a force of 2.8 N has an acceleration of 3.6 m/s^2. What is the mass of the object?

16. What is the acceleration a 1.4-kg object falling through the air if the force of air resistance on the object is 2.5 N?

17. Name three ways you could accelerate if you were riding a bicycle.

Part 3 Open Ended

Record your answers on a sheet of paper.

18. When astronauts orbit Earth, they float inside the spaceship because of weightlessness. Explain this effect.

19. Describe how satellites are able to remain in orbit around Earth.

Use the figure below to answer questions 20 and 21.

20. The figure above shows the path a ball thrown into the air follows. What causes the ball to move along a curved path?

21. What effect would throwing the ball harder have on the ball's path? Explain.

22. How does Newton's second law determine the motion of a book as you push it across a desktop?

23. A heavy box sits on a sidewalk. If you push against the box, the box moves in the direction of the force. If the box is replaced with a ball of the same mass, and you push with the same force against the ball, will it have the same acceleration as the box? Explain.

24. According to Newton's third law of motion, a rock sitting on the ground pushes against the ground, and the ground pushes back against the rock with an equal force. Explain why this force doesn't cause the rock to accelerate upward from the ground according to Newton's second law.

chapter 3

Forces and Fluids

The BIG Idea
A fluid exerts forces on an object in the fluid.

SECTION 1
Pressure
Main Idea The pressure exerted by a fluid increases as depth increases.

SECTION 2
Why do objects float?
Main Idea A fluid exerts an upward buoyant force on any object placed in the fluid.

SECTION 3
Doing Work with Fluids
Main Idea Fluids can be made to exert forces that do useful work.

A Very Fluid Situation

This swimmer seems to be defying gravity as he skims over the water. Even though the swimmer floats, a rock with the same mass would sink to the bottom of the pool. What forces cause something to float? Why does a swimmer float, while a rock sinks?

Science Journal Compare and contrast five objects that float with five objects that sink.

Start-Up Activities

Forces Exerted by Air

When you are lying down, something is pushing down on you with a force equal to the weight of several small cars. What is the substance that is applying all this pressure on you? It's air, a fluid that exerts forces on everything it is in contact with.

1. Suck water into a straw. Try to keep the straw completely filled with water.
2. Quickly cap the top of the straw with your finger and observe what happens to the water.
3. Release your finger from the top of the straw for an instant and replace it as quickly as possible. Observe what happens to the water.
4. Release your finger from the top of the straw and observe.
5. **Think Critically** Write a paragraph describing your observations of the water in the straw. When were the forces acting on the water in the straw balanced and when were they unbalanced?

Fluids Make the following Foldable to compare and contrast the characteristics of two types of fluids—liquids and gases.

STEP 1 Fold one sheet of paper lengthwise.

STEP 2 Fold into thirds.

STEP 3 Unfold and draw overlapping ovals. Cut the top sheet along the folds.

STEP 4 Label the ovals as shown.

Construct a Venn Diagram As you read the chapter, list the characteristics of liquids under the left tab, those characteristics of gases under the right tab, and those characteristics common to both under the middle tab.

Preview this chapter's content and activities at
bookm.msscience.com

M ◆ 65

Get Ready to Read

Make Inferences

① Learn It! When you make inferences, you draw conclusions that are not directly stated in the text. This means you "read between the lines." You interpret clues and draw upon prior knowledge. Authors rely on a reader's ability to infer because all the details are not always given.

② Practice It! Read the excerpt below and pay attention to highlighted words as you make inferences. Use this Think-Through chart to help you make inferences.

The buoyant force pushes an object in a fluid upward, but **gravity pulls the object downward**. If the **weight of the object** is greater than the buoyant force, the **net force** on the object is downward and it sinks.

— *from page 75*

Text	Question	Inferences
Gravity pulls the object downward	What is gravity?	A downward force on the object?
Weight of the object	Is weight a force?	Weight is downward pull due to gravity?
Net force	What is the net force?	Combination of all forces on an object?

③ Apply It! As you read this chapter, practice your skill at making inferences by making connections and asking questions.

Target Your Reading

Reading Tip
Sometimes you make inferences by using other reading skills, such as questioning and predicting.

Use this to focus on the main ideas as you read the chapter.

1 Before you read the chapter, respond to the statements below on your worksheet or on a numbered sheet of paper.
- Write an **A** if you **agree** with the statement.
- Write a **D** if you **disagree** with the statement.

2 After you read the chapter, look back to this page to see if you've changed your mind about any of the statements.
- If any of your answers changed, explain why.
- Change any false statements into true statements.
- Use your revised statements as a study guide.

Before You Read A or D		Statement	After You Read A or D
	1	The pressure due to a force depends on the area over which a force is exerted.	
	2	Fluids are substances that can flow.	
	3	The pressure exerted by water depends on the shape of its container.	
	4	The pressure in a fluid decreases with depth.	
	5	A rock sinks in water because the rock's weight is less than the upward force exerted by the water.	
	6	A metal boat floats in water because the density of the boat is less than the density of water.	
	7	Squeezing on a closed plastic water bottle causes the pressure in the water to decrease.	
	8	The pressure exerted by fluid decreases as the speed of the fluid increases.	
	9	The lift on a wing depends on the size of the wing.	

Science Online
Print out a worksheet of this page at bookm.msscience.com

M ◆ 66 B

section 1

Pressure

as you read

What You'll Learn
- **Define and calculate** pressure.
- **Model** how pressure varies in a fluid.

Why It's Important
Some of the processes that help keep you alive, such as inhaling and exhaling, depend on differences in pressure.

Review Vocabulary
weight: on Earth, the gravitational force between an object and Earth

New Vocabulary
- pressure
- fluid

What is pressure?

What happens when you walk in deep, soft snow or dry sand? Your feet sink into the snow or sand and walking can be difficult. If you rode a bicycle with narrow tires over these surfaces, the tires would sink even deeper than your feet.

How deep you sink depends on your weight as well as the area over which you make contact with the sand or snow. Like the person in **Figure 1**, when you stand on two feet, you make contact with the sand over the area covered by your feet. However, if you were to stand on a large piece of wood, your weight would be distributed over the area covered by the wood.

In both cases, your weight exerted a downward force on the sand. What changed was the area of contact between you and the sand. By changing the area of contact, you changed the pressure you exerted on the sand due to your weight. **Pressure** is the force per unit area that is applied on the surface of an object. When you stood on the board, the area of contact increased, so that the same force was applied over a larger area. As a result, the pressure that was exerted on the sand decreased and you didn't sink as deep.

Figure 1 When your weight is distributed over a larger area, the pressure you exert on the sand decreases.

Calculating Pressure What would happen to the pressure exerted by your feet if your weight increased? You might expect that you would sink deeper in the sand, so the pressure also would increase. Pressure increases if the force applied increases, and decreases if the area of contact increases. Pressure can be calculated from this formula.

Pressure Equation

$$\text{Pressure (in pascals)} = \frac{\text{force (in newtons)}}{\text{area (in meters squared)}}$$

$$P = \frac{F}{A}$$

The unit of pressure in the SI system is the pascal, abbreviated Pa. One pascal is equal to a force of 1 N applied over an area of 1 m², or 1 Pa = 1 N/m². The weight of a dollar bill resting completely flat on a table exerts a pressure of about 1 Pa on the table. Because 1 Pa is a small unit of pressure, pressure sometimes is expressed in units of kPa, which is 1,000 Pa.

Topic: Snowshoes
Visit bookm.msscience.com for Web links to information about the history and use of snowshoes. These devices have been used for centuries in cold, snowy climates.

Activity Use simple materials, such as pipe cleaners, string, or paper, to make a model of a snowshoe.

Applying Math — Solve One-Step Equations

CALCULATING PRESSURE A water glass sitting on a table weighs 4 N. The bottom of the water glass has a surface area of 0.003 m². Calculate the pressure the water glass exerts on the table.

Solution

1. This is what you know:
   - force: $F = 4$ N
   - area: $A = 0.003$ m²

2. This is what you need to find out:
   - pressure: $P = ?$ Pa

3. This is the procedure you need to use:

   Substitute the known values for force and area into the pressure equation and calculate the pressure:

   $$P = \frac{F}{A} = \frac{(4 \text{ N})}{(0.003 \text{ m}^2)}$$
   $$= 1{,}333 \text{ N/m}^2 = 1{,}333 \text{ Pa}$$

4. Check your answer:

   Multiply pressure by the given area. You should get the force that was given.

Practice Problems

1. A student weighs 600 N. The student's shoes are in contact with the floor over a surface area of 0.012 m². Calculate the pressure exerted by the student on the floor.

2. A box that weighs 250 N is at rest on the floor. If the pressure exerted by the box on the floor is 25,000 Pa, over what area is the box in contact with the floor?

 For more practice, visit bookm.msscience.com/math_practice

Mini LAB

Interpreting Footprints

Procedure
1. Go outside to **an area of dirt, sand, or snow** where you can make footprints. Smooth the surface.
2. Make tracks in several different ways. Possible choices include walking forward, walking backward, running, jumping a short or long distance, walking carrying a load, and tiptoeing.

Analysis
1. Measure the depth of each type of track at two points: the ball of the foot and the heel. Compare the depths of the different tracks.
2. The depth of the track corresponds to the pressure on the ground. In your **Science Journal,** explain how different means of motion put pressure on different parts of the sole.
3. Have one person make a track while the other looks away. Then have the second person determine what the motion was.

Pressure and Weight To calculate the pressure that is exerted on a surface, you need to know the force and the area over which it is applied. Sometimes the force that is exerted is the weight of an object, such as when you are standing on sand, snow, or a floor. Suppose you are holding a 2-kg book in the palm of your hand. To find out how much pressure is being exerted on your hand, you first must know the force that the book is exerting on your hand—its weight.

$$\text{Weight} = \text{mass} \times \text{acceleration due to gravity}$$
$$W = (2 \text{ kg}) \times (9.8 \text{ m/s}^2)$$
$$W = 19.6 \text{ N}$$

If the area of contact between your hand and the book is 0.003 m², the pressure that is exerted on your hand by the book is:

$$P = \frac{F}{A}$$
$$P = \frac{(19.6 \text{ N})}{(0.003 \text{ m}^2)}$$
$$P = 6{,}533 \text{ Pa} = 6.53 \text{ kPa}$$

Pressure and Area One way to change the pressure that is exerted on an object is to change the area over which the force is applied. Imagine trying to drive a nail into a piece of wood, as shown in **Figure 2**. Why is the tip of a nail pointed instead of flat? When you hit the nail with a hammer, the force you apply is transmitted through the nail from the head to the tip. The tip of the nail comes to a point and is in contact with the wood over a small area. Because the contact area is so small, the pressure that is exerted by the nail on the wood is large—large enough to push the wood fibers apart. This allows the nail to move downward into the wood.

Figure 2 The force applied to the head of the nail by the hammer is the same as the force that the tip of the nail applies to the wood. However, because the area of the tip is small, the pressure applied to the wood is large.

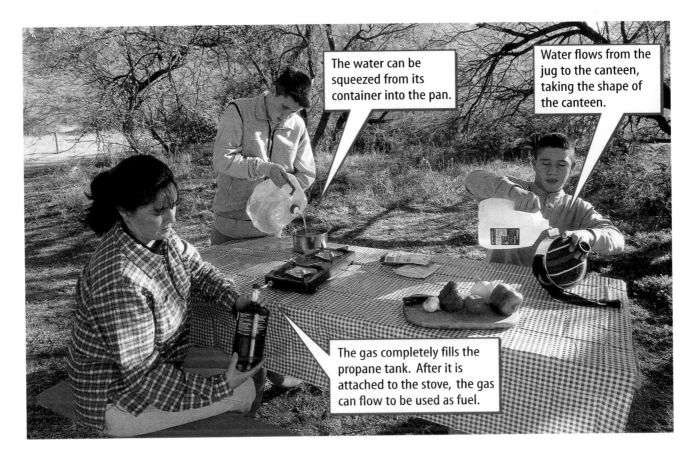

Fluids

What do the substances in **Figure 3** have in common? Each takes the shape of its container and can flow from one place to another. A **fluid** is any substance that has no definite shape and has the ability to flow. You might think of a fluid as being a liquid, such as water or motor oil. But gases are also fluids. When you are outside on a windy day, you can feel the air flowing past you. Because air can flow and has no definite shape, air is a fluid. Gases, liquids, and the state of matter called plasma, which is found in the Sun and other stars, are fluids and can flow.

Pressure in a Fluid

Suppose you placed an empty glass on a table. The weight of the glass exerts pressure on the table. If you fill the glass with water, the weight of the water and glass together exert a force on the table. So the pressure exerted on the table increases.

Because the water has weight, the water itself also exerts pressure on the bottom of the glass. This pressure is the weight of the water divided by the area of the glass bottom. If you pour more water into the glass, the height of the water in the glass increases and the weight of the water increases. As a result, the pressure exerted by the water increases.

Figure 3 Fluids all have the ability to flow and take the shape of their containers.
Classify *What are some other examples of fluids?*

Plasma The Sun is a star with a core temperature of about 16 million°C. At this temperature, the particles in the Sun move at tremendous speeds, crashing into each other in violent collisions that tear atoms apart. As a result, the Sun is made of a type of fluid called a plasma. A plasma is a gas made of electrically charged particles.

SECTION 1 Pressure **M ◆ 69**

Figure 4 Even though each graduated cylinder contains the same volume of water, the pressure exerted by the higher column of water is greater.
Infer *how the pressure exerted by a water column would change as the column becomes narrower.*

Pressure and Fluid Height Suppose you poured the same amount of water into a small and a large graduated cylinder, as shown in **Figure 4.** Notice that the height of the water in the small cylinder is greater than in the large cylinder. Is the water pressure the same at the bottom of each cylinder? The weight of the water in each cylinder is the same, but the contact area at the bottom of the small cylinder is smaller. Therefore, the pressure is greater at the bottom of the small cylinder.

The height of the water can increase if more water is added to a container or if the same amount of water is added to a narrower container. In either case, when the height of the fluid is greater, the pressure at the bottom of the container is greater. This is always true for any fluid or any container. The greater the height of a fluid above a surface, the greater the pressure exerted by the fluid on that surface. The pressure exerted at the bottom of a container doesn't depend on the shape of the container, but only on the height of the fluid above the bottom, as **Figure 5** shows.

Pressure Increases with Depth If you swim underwater, you might notice that you can feel pressure in your ears. As you go deeper, you can feel this pressure increase. This pressure is exerted by the weight of the water above you. As you go deeper in a fluid, the height of the fluid above you increases. As the height of the fluid above you increases, the weight of the fluid above you also increases. As a result, the pressure exerted by the fluid increases with depth.

Figure 5 Pressure depends only on the height of the fluid above a surface, not on the shape of the container. The pressure at the bottom of each section of the tube is the same.

Pressure in All Directions If the pressure that is exerted by a fluid is due to the weight of the fluid, is the pressure in a fluid exerted only downward? **Figure 6** shows a small, solid cube in a fluid. The fluid exerts a pressure on each side of this cube, not just on the top. The pressure on each side is perpendicular to the surface, and the amount of pressure depends only on the depth in the fluid. As shown in **Figure 6,** this is true for any object in a fluid, no matter how complicated the shape. The pressure at any point on the object is perpendicular to the surface of the object at that point.

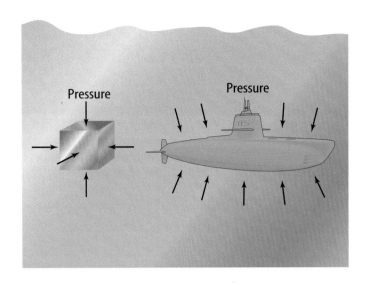

Figure 6 The pressure on all objects in a fluid is exerted on all sides, perpendicular to the surface of the object, no matter what its shape.

Reading Check *In what direction is pressure exerted by a fluid on a surface?*

Atmospheric Pressure

Even though you don't feel it, you are surrounded by a fluid that exerts pressure on you constantly. That fluid is the atmosphere. The atmosphere at Earth's surface is only about one-thousandth as dense as water. However, the thickness of the atmosphere is large enough to exert a large pressure on objects at Earth's surface. For example, look at **Figure 7.** When you are sitting down, the force pushing down on your body due to atmospheric pressure can be equal to the weight of several small cars. Atmospheric pressure is approximately 100,000 Pa at sea level. This means that the weight of Earth's atmosphere exerts about 100,000 N of force over every square meter on Earth.

Why doesn't this pressure cause you to be crushed? Your body is filled with fluids such as blood that also exert pressure. The pressure exerted outward by the fluids inside your body balances the pressure exerted by the atmosphere.

Going Higher As you go higher in the atmosphere, atmospheric pressure decreases as the amount of air above you decreases. The same is true in an ocean, lake, or pond. The water pressure is highest at the ocean floor and decreases as you go upward. The changes in pressure at varying heights in the atmosphere and depths in the ocean are illustrated in **Figure 8.**

Figure 7 Atmospheric pressure on your body is a result of the weight of the atmosphere exerting force on your body.
Infer *Why don't you feel the pressure exerted by the atmosphere?*

SECTION 1 Pressure **M** ◆ **71**

NATIONAL GEOGRAPHIC VISUALIZING PRESSURE AT VARYING ELEVATIONS

Figure 8

No matter where you are on Earth, you're under pressure. Air and water have weight and therefore exert pressure on your body. The amount of pressure depends on your location above or below sea level and how much air or water—or both—are exerting force on you.

▲ **HIGH ELEVATION** With increasing elevation, the amount of air above you decreases, and so does air pressure. At the 8,850-m summit of Mt. Everest, air pressure is a mere 33 kPa—about one third of the air pressure at sea level.

▲ **SEA LEVEL** Air pressure is pressure exerted by the weight of the atmosphere above you. At sea level the atmosphere exerts a pressure of about 100,000 N on every square meter of area. Called one atmosphere (atm), this pressure is also equal to 100 kPa.

▶ **REEF LEVEL** When you descend below the sea surface, pressure increases by about 1 atm every 10 meters. At 20 meters depth, you'd experience 2 atm of water pressure and 1 atm of air pressure, a total of 3 atm of pressure on your body.

▶ **VERY LOW ELEVATION** The deeper you dive, the greater the pressure. The water pressure on a submersible at a depth of 2,200 m is about 220 times greater than atmospheric pressure at sea level.

Barometer An instrument called a barometer is used to measure atmospheric pressure. A barometer has something in common with a drinking straw. When you drink through a straw, it seems like you pull your drink up through the straw. But actually, atmospheric pressure pushes your drink up the straw. By removing air from the straw, you reduce the air pressure in the straw. Meanwhile, the atmosphere is pushing down on the surface of your drink. When you pull the air from the straw, the pressure in the straw is less than the pressure pushing down on the liquid, so atmospheric pressure pushes the drink up the straw.

One type of barometer works in a similar way, as shown in **Figure 9.** The space at the top of the tube is a vacuum. Atmospheric pressure pushes liquid up a tube. The liquid reaches a height where the pressure at the bottom of the column of liquid balances the pressure of the atmosphere. As the atmospheric pressure changes, the force pushing on the surface of the liquid changes. As a result, the height of the liquid in the tube increases as the atmospheric pressure increases.

Figure 9 In this type of barometer, the height of the liquid column increases as the atmospheric pressure increases.

section 1 review

Summary

Pressure
- Pressure is the force exerted on a unit area of a surface. Pressure can be calculated from this equation:

$$P = \frac{F}{A}$$

- The SI unit for pressure is the pascal, abbreviated Pa.

Pressure in a Fluid
- The pressure exerted by a fluid depends on the depth below the fluid surface.
- The pressure exerted by a fluid on a surface is always perpendicular to the surface.

Atmospheric Pressure
- Earth's atmosphere exerts a pressure of about 100,000 Pa at sea level.
- A barometer is an instrument used to measure atmospheric pressure.

Self Check

1. **Compare** One column of water has twice the diameter as another water column. If the pressure at the bottom of each column is the same, how do the heights of the two columns compare?
2. **Explain** why the height of the liquid column in a barometer changes as atmospheric pressure changes.
3. **Classify** the following as fluids or solids: warm butter, liquid nitrogen, paper, neon gas, ice.
4. **Explain** how the pressure at the bottom of a container depends on the container shape and the fluid height.
5. **Think Critically** Explain how the diameter of a balloon changes as it rises higher in the atmosphere.

Applying Math

6. **Calculate Force** The palm of a person's hand has an area of 0.0135 m². If atmospheric pressure is 100,000 N/m², find the force exerted by the atmosphere on the person's palm.

section 2
Why do objects float?

as you read

What You'll Learn
- **Explain** how the pressure in a fluid produces a buoyant force.
- **Define** density.
- **Explain** floating and sinking using Archimedes' principle.

Why It's Important
Knowing how fluids exert forces helps you understand how boats can float.

Review Vocabulary
Newton's second law of motion: the acceleration of an object is in the direction of the total force and equals the total force divided by the object's mass

New Vocabulary
- buoyant force
- Archimedes' principle
- density

The Buoyant Force

Can you float? Think about the forces that are acting on you as you float motionless on the surface of a pool or lake. You are not moving, so according to Newton's second law of motion, the forces on you must be balanced. Earth's gravity is pulling you downward, so an upward force must be balancing your weight, as shown in **Figure 10**. This force is called the buoyant force. The **buoyant force** is an upward force that is exerted by a fluid on any object in the fluid.

What causes the buoyant force?

The buoyant force is caused by the pressure that is exerted by a fluid on an object in the fluid. **Figure 11** shows a cube-shaped object submerged in a glass of water. The water exerts pressure everywhere over the surface of the object. Recall that the pressure exerted by a fluid has two properties. One is that the direction of the pressure on a surface is always perpendicular to the surface. The other is that the pressure exerted by a fluid increases as you go deeper into the fluid.

Figure 10 When you float, the forces on you are balanced. Gravity pulls you downward and is balanced by the buoyant force pushing you upward.
Infer *What is the acceleration of the person shown here?*

Buoyant Force and Unbalanced Pressure The pressure that is exerted by the water on the cube is shown in **Figure 11.** The bottom of the cube is deeper in the water. Therefore, the pressure that is exerted by the water at the bottom of the cube is greater than it is at the top of the cube. The higher pressure near the bottom means that the water exerts an upward force on the bottom of the cube that is greater than the downward force that is exerted at the top of the cube. As a result, the force that is exerted on the cube due to water pressure is not balanced, and a net upward force is acting on the cube due to the pressure of the water. This upward force is the buoyant force. A buoyant force acts on all objects that are placed in a fluid, whether they are floating or sinking.

Reading Check *When does the buoyant force act on an object?*

Figure 11 The pressure exerted on the bottom of the cube is greater than the pressure on the top. The fluid exerts a net upward force on the cube.

Sinking and Floating

If you drop a stone into a pool of water, it sinks. But if you toss a twig on the water, it floats. An upward buoyant force acts on the twig and the stone, so why does one float and one sink?

The buoyant force pushes an object in a fluid upward, but gravity pulls the object downward. If the weight of the object is greater than the buoyant force, the net force on the object is downward and it sinks. If the buoyant force is equal to the object's weight, the forces are balanced and the object floats. As shown in **Figure 12,** the fish floats because the buoyant force on it balances its weight. The rocks sink because the buoyant force acting on them is not large enough to balance their weight.

Figure 12 The weight of a rock is more than the buoyant force exerted by the water, so it sinks to the bottom.
Infer *Why do the fish float?*

SECTION 2 Why do objects float?

If the foil is folded up, its surface area is small. The buoyant force is less than the foil's weight and it sinks.

If the foil is unfolded, its surface area is larger. The buoyant force is equal to its weight and it floats.

Figure 13 The buoyant force on a piece of aluminum foil increases as the surface area of the foil increases.

Figure 14 The hull of this oil tanker has a large surface area in contact with the water. As a result, the buoyant force is so large that the ship floats.

Changing the Buoyant Force

Whether an object sinks or floats depends on whether the buoyant force is smaller than its weight. The weight of an object depends only on the object's mass, which is the amount of matter the object contains. The weight does not change if the shape of the object changes. A piece of modeling clay contains the same amount of matter whether it's squeezed into a ball or pressed flat.

Buoyant Force and Shape Buoyant force does depend on the shape of the object. The fluid exerts upward pressure on the entire lower surface of the object that is in contact with the fluid. If this surface is made larger, then more upward pressure is exerted on the object and the buoyant force is greater. **Figure 13** shows how a piece of aluminum can be made to float. If the aluminum is crumpled, the buoyant force is less than the weight, so the aluminum sinks. When the aluminum is flattened into a thin, curved sheet, the buoyant force is large enough that the sheet floats. This is how large, metal ships, like the one in **Figure 14,** are able to float. The metal is formed into a curved sheet that is the hull of the ship. The contact area of the hull with the water is much greater than if the metal were a solid block. As a result, the buoyant force on the hull is greater than it would be on a metal block.

76 ◆ **M CHAPTER 3** Forces and Fluids

The Buoyant Force Doesn't Change with Depth

Suppose you drop a steel cube into the ocean. You might think that the cube would sink only to a depth where the buoyant force on the cube balances its weight. However, the steel sinks to the bottom, no matter how deep the ocean is.

The buoyant force on the cube is the difference between the downward force due to the water pressure on the top of the cube and the upward force due to water pressure on the bottom of the cube. **Figure 15** shows that when the cube is deeper, the pressure on the top surface increases, but the pressure on the bottom surface also increases by the same amount. As a result, the difference between the forces on the top and bottom surfaces is the same, no matter how deep the cube is submerged. The buoyant force on the submerged cube is the same at any depth.

Archimedes' Principle

A way of determining the buoyant force was given by the ancient Greek mathematician Archimedes(ar kuh MEE deez) more than 2,200 years ago. According to **Archimedes' principle,** the buoyant force on an object is equal to the weight of the fluid it displaces.

To understand Archimedes' principle, think about what happens if you drop an ice cube in a glass of water that's filled to the top. The ice cube takes the place of some of the water and causes this water to overflow, as shown in **Figure 16.** Another way to say this is that the ice cube displaced water that was in the glass.

Suppose you caught all the overflow water and weighed it. According to Archimedes' principle, the weight of the overflow, or displaced water, would be equal to the buoyant force on the ice cube. Because the ice cube is floating, the buoyant force is balanced by the weight of the ice cube. So the weight of the water that is displaced, or the buoyant force, is equal to the weight of the ice cube.

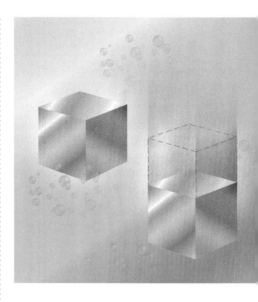

Figure 15 Because the cube on the right is deeper, the pressure on its upper surface is increased due to the weight of the water inside the dashed lines. The pressure on the bottom surface also increases by this amount.
Explain *how the buoyant force on the cube would change if it moved only to the left or right.*

Figure 16 The buoyant force exerted on this ice cube is equal to the weight of the water displaced by the ice cube.

Naval Architect Naval architects design the ships and submarines for the U.S. Naval Fleet, Coast Guard, and Military Sealift Command. Naval architects need math, science, and English skills for designing and communicating design ideas to others.

Density Archimedes' principle leads to a way of determining whether an object placed in a fluid will float or sink. The answer depends on comparing the density of the fluid and the density of the object. The **density** of a fluid or an object is the mass of the object divided by the volume it occupies. Density can be calculated by the following formula:

Density Equation
$$\text{density (in g/cm}^3) = \frac{\text{mass (in g)}}{\text{volume (in cm}^3)}$$
$$D = \frac{m}{V}$$

For example, water has a density of 1.0 g/cm³. The mass of any volume of a substance can be calculated by multiplying both sides of the above equation by volume. This gives the equation

$$\text{mass} = \text{density} \times \text{volume}$$

Then if the density and volume are known, the mass of the material can be calculated.

Applying Science

Layering Liquids

The density of an object or substance determines whether it will sink or float in a fluid. Just like solid objects, liquids also have different densities. If you pour vegetable oil into water, the oil doesn't mix. Instead, because the density of oil is less than the density of water, the oil floats on top of the water.

Identifying the Problem

In science class, a student is presented with five unknown liquids and their densities. He measures the volume of each and organizes his data into the table at the right. He decides to experiment with these liquids by carefully pouring them, one at a time, into a graduated cylinder.

Liquid Density and Volume			
Liquid	Color	Density (g/cm³)	Volume (cm³)
A	red	2.40	32.0
B	blue	2.90	15.0
C	green	1.20	20.0
D	yellow	0.36	40.0
E	purple	0.78	19.0

Solving the Problem
1. Assuming the liquids don't mix with each other, draw a diagram and label the colors, illustrating how these liquids would look when poured into a graduated cylinder. If 30 cm³ of water were added to the graduated cylinder, explain how your diagram would change.
2. Use the formula for density to calculate the mass of each of the unknown liquids in the chart.

Sinking and Density Suppose you place a copper block with a volume of 1,000 cm³ into a container of water. This block weighs about 88 N. As the block sinks, it displaces water, and an upward buoyant force acts on it. If the block is completely submerged, the volume of water it has displaced is 1,000 cm³—the same as its own volume. This is the maximum amount of water the block can displace. The weight of 1,000 cm³ of water is about 10 N, and this is the maximum buoyant force that can act on the block. This buoyant force is less than the weight of the copper, so the copper block continues to sink.

The copper block and the displaced water had the same volume. Because the copper block had a greater density, the mass of the copper block was greater than the mass of the displaced water. As a result, the copper block weighed more than the displaced water because its density was greater. Any material with a density that is greater than the density of water will weigh more than the water that it displaces, and it will sink. This is true for any object and any fluid. Any object that has a density greater than the density of the fluid it is placed in will sink.

Floating and Density Suppose you place a block of wood with a volume of 1,000 cm³ into a container of water. This block weighs about 7 N. The block starts to sink and displaces water. However, it stops sinking and floats before it is completely submerged, as shown in **Figure 17.** The density of the wood was less than the density of the water. So the wood was able to displace an amount of water equal to its weight before it was completely submerged. It stopped sinking after it had displaced about 700 cm³ of water. That much water has a weight of about 7 N, which is equal to the weight of the block. Any object with a density less than the fluid it is placed in will float.

Figure 17 An object, such as this block of wood, will continue to sink in a fluid until it has displaced an amount of fluid that is equal to its mass. Then the buoyant force equals its weight.

 How can you determine whether an object will float or sink?

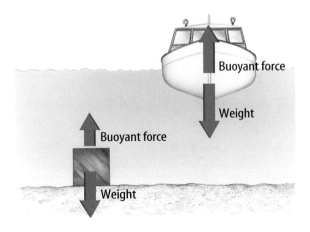

Figure 18 Even though the boat and the cube have the same mass, the boat displaces more water because of its shape. Therefore the boat floats, but the cube sinks.

Boats

Archimedes' principle provides another way to understand why boats that are made of metal can float. Look at **Figure 18.** By making a piece of steel into a boat that occupies a large volume, more water is displaced by the boat than by the piece of steel. According to Archimedes' principle, increasing the weight of the water that is displaced increases the buoyant force. By making the volume of the boat large enough, enough water can be displaced so that the buoyant force is greater than the weight of the steel.

How does the density of the boat compare to the density of the piece of steel? The steel now surrounds a volume that is filled with air that has little mass. The mass of the boat is nearly the same as the mass of the steel, but the volume of the boat is much larger. As a result, the density of the boat is much less than the density of the steel. The boat floats when its volume becomes large enough that its density is less than the density of water.

section 2 review

Summary

The Buoyant Force
- The buoyant force is an upward force that is exerted by a fluid on any object in the fluid.
- The buoyant force is caused by the increase in pressure with depth in a fluid.
- Increasing the surface area in contact with a fluid increases the buoyant force on an object.

Sinking and Floating
- An object sinks when the buoyant force on an object is less than the object's weight.
- An object floats when the buoyant force on an object equals the object's weight.

Archimedes' Principle
- Archimedes' principle states that the buoyant force on a object equals the weight of the fluid the object displaces.
- According to Archimedes' principle, an object will float in a fluid only if the density of the object is less than the density of the fluid.

Self Check

1. **Explain** whether the buoyant force on a submerged object depends on the weight of the object.
2. **Determine** whether an object will float or sink in water if it has a density of 1.5 g/cm^3. Explain.
3. **Compare** the buoyant force on an object when it is partially submerged and when it's completely submerged.
4. **Explain** how the buoyant force acting on an object placed in water can be measured.
5. **Think Critically** A submarine changes its mass by adding or removing seawater from tanks inside the sub. Explain how this can enable the sub to dive or rise to the surface.

Applying Math

6. **Calculate Buoyant Force** A ship displaces 80,000 L of water. One liter of water weighs 9.8 N. What is the buoyant force on the ship?
7. **Density** The density of 14k gold is 13.7 g/cm^3. What is the density of a ring with a mass of 7.21 g and a volume of 0.65 cm^3. Is it made from 14k gold?

Measuring Buoyant Force

The total force on an object in a fluid is the difference between the object's weight and the buoyant force. In this lab, you will measure the buoyant force on an object and compare it to the weight of the water displaced.

Real-World Question

How is the buoyant force related to the weight of the water that an object displaces?

Goals

- **Measure** the buoyant force on an object.
- **Compare** the buoyant force to the weight of the water displaced by the object.

Materials

aluminum pan
spring scale
500-mL beaker
graduated cylinder
funnel
metal object

Safety Precautions

Procedure

1. Place the beaker in the aluminum pan and fill the beaker to the brim with water.
2. Hang the object from the spring scale and record its weight.
3. With the object hanging from the spring scale, completely submerge the object in the water. The object should not be touching the bottom or the sides of the beaker.
4. **Record** the reading on the spring scale while the object is in the water. Calculate the buoyant force by subtracting this reading from the object's weight.
5. Use the funnel to carefully pour the water from the pan into the graduated cylinder. Record the volume of this water in cm³.
6. **Calculate** the weight of the water displaced by multiplying the volume of water by 0.0098 N.

Conclude and Apply

1. **Explain** how the total force on the object changed when it was submerged in water.
2. **Compare** the weight of the water that is displaced with the buoyant force.
3. **Explain** how the buoyant force would change if the object were submersed halfway in water.

Communicating Your Data

Make a poster of an empty ship, a heavily loaded ship, and an overloaded, sinking ship. Explain how Archimedes' principle applies in each case. **For more help, refer to the Science Skill Handbook.**

section 3
Doing Work with Fluids

as you read

What You'll Learn
- **Explain** how forces are transmitted through fluids.
- **Describe** how a hydraulic system increases force.
- **Describe** Bernoulli's principle.

Why It's Important
Fluids can exert forces that lift heavy objects and enable aircraft to fly.

Review Vocabulary
work: the product of the force applied to an object and the distance the object moves in the direction of the force

New Vocabulary
- Pascal's principle
- hydraulic system
- Bernoulli's principle

Using Fluid Forces

You might have watched a hydraulic lift raise a car off the ground. It might surprise you to learn that the force pushing the car upward is being exerted by a fluid. When a huge jetliner soars through the air, a fluid exerts the force that holds it up. Fluids at rest and fluids in motion can be made to exert forces that do useful work, such as pumping water from a well, making cars stop, and carrying people long distances through the air. How are these forces produced by fluids?

Pushing on a Fluid The pressure in a fluid can be increased by pushing on the fluid. Suppose a watertight, movable cover, or piston, is sitting on top of a column of fluid in a container. If you push on the piston, the fluid can't escape past the piston, so the height of the fluid in the container doesn't change. As a result, the piston doesn't move. But now the force exerted on the bottom of the container is the weight of the fluid plus the force pushing the piston down. Because the force exerted by the fluid at the bottom of the container has increased, the pressure exerted by the fluid also has increased. **Figure 19** shows how the force exerted on a brake pedal is transmitted to a fluid.

Figure 19 Because the fluid in this piston can't escape, it transmits the force you apply throughout the fluid.

Pascal's Principle

Suppose you fill a plastic bottle with water and screw the cap back on. If you poke a hole in the bottle near the top, water will leak out of the hole. However, if you squeeze the bottle near the bottom, as shown in **Figure 20,** water will shoot out of the hole. When you squeezed the bottle, you applied a force on the fluid. This increased the pressure in the fluid and pushed the water out of the hole faster.

No matter where you poke the hole in the bottle, squeezing the bottle will cause the water to flow faster out of the hole. The force you exert on the fluid by squeezing has been transmitted to every part of the bottle. This is an example of Pascal's principle. According to **Pascal's principle,** when a force is applied to a fluid in a closed container, the pressure in the fluid increases everywhere by the same amount.

Hydraulic Systems

Pascal's principle is used in building hydraulic systems like the ones used by car lifts. A **hydraulic system** uses a fluid to increase an input force. The fluid enclosed in a hydraulic system transfers pressure from one piston to another. An example is shown in **Figure 21.** An input force that is applied to the small piston increases the pressure in the fluid. This pressure increase is transmitted throughout the fluid and acts on the large piston. The force the fluid exerts on the large piston is the pressure in the fluid times the area of the piston. Because the area of the large piston is greater than the area of the small piston, the output force exerted on the large piston is greater than the input force exerted on the small piston.

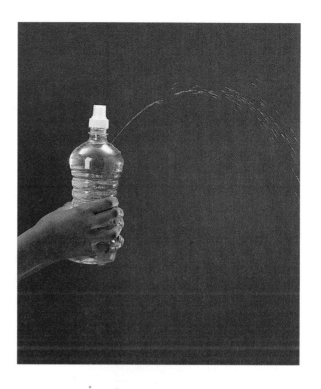

Figure 20 When you squeeze the bottle, the pressure you apply is distributed throughout the fluid, forcing the water out the hole.

Figure 21 A hydraulic system uses Pascal's principle to make the output force applied on the large piston greater than the input force applied on the small piston.
Infer *how the force on the large piston would change if its area increased.*

Science Online

Topic: Hydraulic Systems
Visit bookm.msscience.com for Web links to information about hydraulic systems.

Activity Draw a diagram showing how one of the hydraulic systems that you find works. Share your diagram with the class.

Increasing Force What is the force pushing upward on the larger piston? For example, suppose that the area of the small piston is 1 m^2 and the area of the large piston is 2 m^2. If you push on the small piston with a force of 10 N, the increase in pressure at the bottom of the small piston is

$$\begin{aligned} P &= F/A \\ &= (10 \text{ N})/(1 \text{ m}^2) \\ &= 10 \text{ Pa} \end{aligned}$$

According to Pascal's principle, this increase in pressure is transmitted throughout the fluid. This causes the force exerted by the fluid on the larger piston to increase. The increase in the force on the larger piston can be calculated by multiplying both sides of the above formula by *A*.

$$\begin{aligned} F &= P \times A \\ &= 10 \text{ Pa} \times 2 \text{ m}^2 \\ &= 20 \text{ N} \end{aligned}$$

The force pushing upward on the larger piston is twice as large as the force pushing downward on the smaller piston. What happens if the larger piston increases in size? Look at the calculation above. If the area of the larger piston increases to 5 m^2, the force pushing up on this piston increases to 50 N. So a small force pushing down on the left piston as in **Figure 21** can be made much larger by increasing the size of the piston on the right.

 How does a hydraulic system increase force?

Pressure in a Moving Fluid

What happens to the pressure in a fluid if the fluid is moving? Try the following experiment. Place an empty soda can on the desktop and blow to the right of the can, as shown in **Figure 22**. In which direction will the can move?

When you blow to the right of the can, the can moves to the right, toward the moving air. The air pressure exerted on the right side of the can, where the air is moving, is less than the air pressure on the left side of the can, where the air is not moving. As a result, the force exerted by the air pressure on the left side is greater than the force exerted on the right side, and the can is pushed to the right. What would happen if you blew between two empty cans?

Figure 22 By blowing on one side of the can, you decrease the air pressure on that side. Because the pressure on the opposite side is now greater, the can moves toward the side you're blowing on.

Bernoulli's Principle

The reason for the surprising behavior of the can in **Figure 22** was discovered by the Swiss scientist Daniel Bernoulli in the eighteenth century. It is an example of Bernoulli's principle. According to **Bernoulli's principle,** when the speed of a fluid increases, the pressure exerted by the fluid decreases. When you blew across the side of the can, the pressure exerted by the air on that side of the can decreased because the air was moving faster than it was on the other side. As a result, the can was pushed toward the side you blew across.

Chimneys and Bernoulli's Principle In a fireplace the hotter, less dense air above the fire is pushed upward by the cooler, denser air in the room. Wind outside of the house can increase the rate at which the smoke rises. Look at **Figure 23.** Air moving across the top of the chimney causes the air pressure above the chimney to decrease according to Bernoulli's principle. As a result, more smoke is pushed upward by the higher pressure of the air in the room.

Damage from High Winds You might have seen photographs of people preparing for a hurricane by closing shutters over windows or nailing boards across the outside of windows. In a hurricane, the high winds blowing outside the house cause the pressure outside the house to be less than the pressure inside. This difference in pressure can be large enough to cause windows to be pushed out and to shatter.

Hurricanes and other high winds sometimes can blow roofs from houses. When wind blows across the roof of a house, the pressure outside the roof decreases. If the wind outside is blowing fast enough, the outside pressure can become so low that the roof can be pushed off the house by the higher pressure of the still air inside.

Wings and Flight

You might have placed your hand outside the open window of a moving car and felt the push on it from the air streaming past. If you angled your hand so it tilted upward into the moving air, you would have felt your hand pushed upward. If you increased the tilt of your hand, you felt the upward push increase. You might not have realized it, but your hand was behaving like an airplane wing. The force that lifted your hand was provided by a fluid—the air.

Observing Bernoulli's Principle

Procedure
1. Tie a piece of **string** to the handle of a **plastic spoon**.
2. Turn on a faucet to make a stream of water.
3. Holding the string, bring the spoon close to the stream of water.

Analysis
Use Bernoulli's principle to explain the motion of the spoon.

Figure 23 The air moving past the chimney lowers the air pressure above the chimney. As a result, smoke is forced up the chimney faster than when air above the chimney is still.

Producing Lift How is the upward force, or lift, on an airplane wing produced? A jet engine pushes the plane forward, or a propeller pulls the plane forward. Air flows over the wings as the plane moves. The wings are tilted upward into the airflow, just like your hand was tilted outside the car window. **Figure 24** shows how the tilt of the wing causes air that has flowed over the wing's upper and lower surfaces to be directed downward.

Lift is created by making the air flow downward. To understand this, remember that air is made of different types of molecules. The motion of these molecules is changed only when a force acts on them. When the air is directed downward, a force is being exerted on the air molecules by the wing.

However, according to Newton's third law of motion, for every action force there is an equal but opposite reaction force. The wing exerts a downward action force on the air. So the air must exert an upward reaction force on the wing. This reaction force is the lift that enables paper airplanes and jet airliners to fly.

Airplane Wings Airplanes have different wing shapes, depending on how the airplane is used. The lift on a wing depends on the amount of air that the wing deflects downward and how fast that air is moving. Lift can be increased by increasing the size or surface area of the wing. A larger wing is able to deflect more air downward.

Look at the planes in **Figure 25.** A plane designed to fly at high speeds, such as a jet fighter, can have small wings. A large cargo plane that carries heavy loads needs large wings to provide a great deal of lift. A glider flies at low speeds and uses long wings that have a large surface area to provide the lift it needs.

Reading Check *How can a wing's lift be increased?*

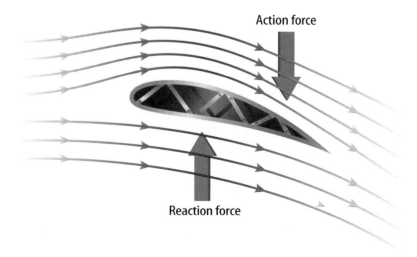

Figure 24 An airplane wing forces air to be directed downward. As a result, the air exerts an upward reaction force on the wing, producing lift.

Figure 25 Different wing shapes are used for different types of planes. Larger wings provide more lift.

 Birds' Wings A bird's wing provides lift in the same way that an airplane wing does. The wings also act as propellers that pull the bird forward when it flaps its wings up and down. Bird wings also have different shapes depending on the type of flight. Seabirds have long, narrow wings, like the wings of a glider, that help them glide long distances. Forest and field birds, such as pheasants, have short, rounded wings that enable them to take off quickly and make sharp turns. Swallows, swifts, and falcons, which fly at high speeds, have small, narrow, tapered wings like those on a jet fighter.

section 3 review

Summary

Pascal's Principle
- Pascal's principle states that when a force is applied to a fluid in a closed container, the pressure in the fluid increases by the same amount everywhere in the fluid.
- Hydraulic systems use Pascal's principle to produce an output force that is greater than an applied input force.
- Increasing the surface area in contact with a fluid increases the buoyant force on an object.

Bernoulli's Principle
- Bernoulli's principle states that when the speed of a fluid increases, the pressure exerted by the fluid decreases.

Wings and Flight
- An airplane wing exerts a force on air and deflects it downward. By Newton's third law, the air exerts an upward reaction force.

Self Check

1. **Explain** why making an airplane wing larger enables the wing to produce more lift.
2. **Infer** If you squeeze a plastic water-filled bottle, where is the pressure change in the water the greatest?
3. **Explain** Use Bernoulli's principle to explain why a car passing a truck tends to be pushed toward the truck.
4. **Infer** why a sheet of paper rises when you blow across the top of the paper.
5. **Think Critically** Explain why the following statement is false: In a hydraulic system, because the increase in the pressure on both pistons is the same, the increase in the force on both pistons is the same.

Applying Math

6. **Calculate Force** The small piston of a hydraulic lift, has an area of 0.01 m^2. If a force of 250 N is applied to the small piston, find the force on the large piston if it has an area of 0.05 m^2.

LAB Use the Internet

Barometric Pressure and Weather

Goals
- **Collect** barometric pressure and other weather data.
- **Compare** barometric pressure to weather conditions.
- **Predict** weather patterns based on barometric pressure, wind speed and direction, and visual conditions.

Data Source

Visit bookm.msscience.com/internet_lab for more information about barometric pressure, weather information, and data collected by other students.

Real-World Question
What is the current barometric pressure where you are? How would you describe the weather today where you are? What is the weather like in the region to the west of you? To the east of you? What will your weather be like tomorrow? The atmosphere is a fluid and flows from one place to another as weather patterns change. Changing weather conditions also cause the atmospheric pressure to change. By collecting barometric pressure data and observing weather conditions, you will be able to make a prediction about the next day's weather.

Make a Plan
1. Visit the Web site on the left for links to information about weather in the region where you live.
2. Find and record the current barometric pressure and whether the pressure is rising, falling, or remaining steady. Also record the wind speed and direction.
3. **Observe and record** other weather conditions, such as whether rain is falling, the Sun is shining, or the sky is cloudy.
4. Based on the data you collect and your observations, predict what you think tomorrow's weather will be. Record your prediction.
5. Repeat the data collection and observation for a total of five days.

88 ◆ **M CHAPTER 3** Forces and Fluids

Using Scientific Methods

Barometric Pressure Weather Data	
Location of weather station	
Barometric pressure	
Status of barometric pressure	
Wind speed	Do not write in
Wind direction	this book.
Current weather conditions	
Predictions of tomorrow's weather conditions	

● Follow Your Plan

1. Make sure your teacher approves your plan before you start.
2. Visit the link below to post your data.

● Analyze Your Data

1. **Analyze** Look at your data. What was the weather the day after the barometric pressure increased? The day after the barometric pressure decreased? The day after the barometric pressure was steady?
2. **Draw Conclusions** How accurate were your weather predictions?

● Conclude and Apply

1. **Infer** What is the weather to the west of you today? How will that affect the weather in your area tomorrow?
2. **Compare** What was the weather to the east of you today? How does that compare to the weather in your area yesterday?
3. **Evaluate** How does increasing, decreasing, or steady barometric pressure affect the weather?

Communicating Your Data

Find this lab using the link below. Use the data on the Web site to predict the weather where you are two days from now.

bookm.msscience.com/internet_lab

LAB **M** ◆ 89

Science and Language Arts

"Hurricane"
by John Balaban

Near dawn our old live oak sagged over
then crashed on the tool shed
rocketing off rakes paintcans flower pots.

All night, rain slashed the shutters until
it finally quit and day arrived in queer light,
silence, and ozoned air. Then voices calling

as neighbors crept out to see the snapped trees,
leaf mash and lawn chairs driven in heaps
with roof bits, siding, sodden birds, dead snakes.

For days, bulldozers clanked by our houses
in sickening August heat as heavy cranes
scraped the rotting tonnage from the streets.

Then our friend Elling drove in from Sarasota
in his old . . . van packed with candles, with
dog food, cat food, flashlights and batteries . . . .

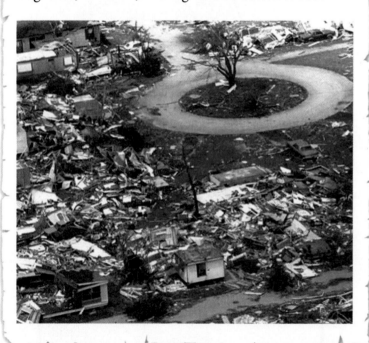

Understanding Literature

Sense Impressions In this poem, John Balaban uses sense impressions to place the reader directly into the poem's environment. For example, the words *rotting tonnage* evoke the sense of smell. Give examples of other sense impressions mentioned.

Respond to the Reading

1. What kinds of damage did the hurricane cause?
2. Why do you think the poet felt relief when his friend, Elling, arrived?
3. **Linking Science and Writing** Write a poem describing a natural phenomenon involving forces and fluids. Use words that evoke at least one of the five sense impressions.

 In the poem, bits of roofs and siding from houses are part of the debris that is everywhere in heaps. According to Bernoulli's principle, the high winds in a hurricane blowing past a house causes the air pressure outside the house to be less than the air pressure inside. In some cases, the forces exerted by the air inside causes the roof to be pushed off the house, and the walls to be blown outward.

chapter 3 Study Guide

Reviewing Main Ideas

Section 1 Pressure

1. Pressure equals force divided by area.
2. Liquids and gases are fluids that flow.
3. Pressure increases with depth and decreases with elevation in a fluid.
4. The pressure exerted by a fluid on a surface is always perpendicular to the surface.

Section 2 Why do objects float?

1. A buoyant force is an upward force exerted on all objects placed in a fluid.
2. The buoyant force depends on the shape of the object.
3. According to Archimedes' principle, the buoyant force on the object is equal to the weight of the fluid displaced by the object.
4. An object floats when the buoyant force exerted by the fluid is equal to the object's weight.
5. An object will float if it is less dense than the fluid it is placed in.

Section 3 Doing Work with Fluids

1. Pascal's principle states that the pressure applied at any point to a confined fluid is transmitted unchanged throughout the fluid.
2. Bernoulli's principle states that when the velocity of a fluid increases, the pressure exerted by the fluid decreases.
3. A wing provides lift by forcing air downward.

Visualizing Main Ideas

Copy and complete the following table.

Relationships Among Forces and Fluids		
Idea	**What does it relate?**	**How?**
Density	mass and volume	
Pressure		force/area
Archimedes' principle	buoyant force and weight of fluid that is displaced	
Bernoulli's principle		velocity increases, pressure decreases
Pascal's principle	pressure applied to enclosed fluid at one point and pressure at other points in a fluid	

chapter 3 Review

Using Vocabulary

Archimedes' principle p. 77
Bernoulli's principle p. 85
buoyant force p. 74
density p. 78
fluid p. 69
hydraulic system p. 83
Pascal's principle p. 83
pressure p. 66

Answer each of the following questions using complete sentences that include vocabulary from the list above.

1. How would you describe a substance that can flow?
2. When the area over which a force is applied decreases, what increases?
3. What principle relates the weight of displaced fluid to the buoyant force?
4. How is a fluid used to lift heavy objects?
5. If you increase an object's mass but not its volume, what have you changed?
6. How is a log able to float in a river?
7. What principle explains why hurricanes can blow the roof off of a house?

Checking Concepts

Choose the word or phrase that best answers the question.

8. Which always equals the weight of the fluid displaced by an object?
   A) the weight of the object
   B) the force of gravity on the object
   C) the buoyant force on the object
   D) the net force on the object

9. What is the net force on a rock that weighs 500 N if the weight of the water it displaces is 300 N?
   A) 200 N downward   C) 800 N downward
   B) 200 N upward   D) 800 N upward

10. The pressure exerted by a fluid on a surface is always in which direction?
    A) upward
    B) downward
    C) parallel to the surface
    D) perpendicular to the surface

Use the photo below to answer question 11.

11. Each graduated cylinder contains the same amount of water. Which of the following statements is true?
    A) The pressure is greater at the bottom of the large cylinder.
    B) The pressure is greater at the bottom of the small cylinder.
    C) The pressure is equal at the bottom of both cylinders.
    D) There is zero pressure at the bottom of both cylinders.

12. Which would increase the lift provided by an airplane wing?
    A) decreasing the volume of the wing
    B) increasing the area of the wing
    C) decreasing the length of the wing
    D) increasing the mass of the wing

13. An airplane wing produces lift by forcing air in which direction?
    A) upward   C) under the wing
    B) downward   D) over the wing

Chapter 3 Review

Thinking Critically

14. **Explain** A sandbag is dropped from a hot-air balloon and the balloon rises. Explain why this happens.

15. **Determine** whether or not this statement is true: Heavy objects sink, and light objects float. Explain your answer.

16. **Explain** why a leaking boat sinks.

17. **Explain** why the direction of the buoyant force on a submerged cube is upward and not left or right.

18. **Recognizing Cause and Effect** A steel tank and a balloon are the same size and contain the same amount of helium. Explain why the balloon rises and the steel tank doesn't.

19. **Make and Use Graphs** Graph the pressure exerted by a 75-kg person wearing different shoes with areas of 0.01 m^2, 0.02 m^2, 0.03 m^2, 0.04 m^2, and 0.05 m^2. Plot pressure on the vertical axis and area on the horizontal axis.

20. **Explain** why it is easier to lift an object that is underwater, than it is to lift the object when it is out of the water.

21. **Infer** Two objects with identical shapes are placed in water. One object floats and the other object sinks. Infer the difference between the two objects that causes one to sink and the other to float.

22. **Compare** Two containers with different diameters are filled with water to the same height. Compare the force exerted by the fluid on the bottom of the two containers.

Performance Activities

23. **Oral Presentation** Research the different wing designs in birds or aircraft. Present your results to the class.

24. **Experiment** Partially fill a plastic dropper with water until it floats just below the surface of the water in a bowl or beaker. Place the dropper inside a plastic bottle, fill the bottle with water, and seal the top. Now squeeze the bottle. What happens to the water pressure in the bottle? How does the water level in the dropper change? How does the density of the dropper change? Use your answers to these questions to explain how the dropper moves when you squeeze the bottle.

Applying Math

25. **Buoyant Force** A rock is attached to a spring scale that reads 10 N. If the rock is submerged in water, the scale reads 6 N. What is the buoyant force on the submerged rock?

26. **Hydraulic Force** A hydraulic lift with a large piston area of 0.04 m^2 exerts a force of 5,000 N. If the smaller piston has an area of 0.01 m^2, what is the force on it?

Use the table below to answer questions 27 and 28.

Material Density	
Substance	Density (g/cm^3)
Ice	0.92
Lead	11.34
Balsa wood	0.12
Sugar	1.59

27. **Density** Classify which of the above substances will and will not float in water.

28. **Volume** Find the volumes of each material, if each has a mass of 25 g.

29. **Pressure** What is the pressure due to a force of 100 N on an area 4 m^2?

30. **Pressure** A bottle of lemonade sitting on a table weighs 6 N. The bottom of the bottle has a surface area of 0.025 m^2. Calculate the pressure the bottle of lemonade exerts on the table.

Part 1 Multiple Choice

Record your answers on the answer sheet provided by your teacher or on a sheet of paper.

1. A force of 15 N is exerted on an area of 0.1 m². What is the pressure?
   A. 150 N
   B. 150 Pa
   C. 1.5 Pa
   D. 0.007 Pa

Use the illustration below to answer questions 2 and 3.

2. Which statement is TRUE?
   A. The contact area between the board and the nail tip is large.
   B. The contact area between the nail tip and the board is greater than the contact area between the nail head and the hammer head.
   C. The nail exerts no pressure on the board because its weight is so small.
   D. The pressure exerted by the nail is large because the contact area between the nail tip and the board is small.

3. Which increases the pressure exerted by the nail tip on the board?
   A. increasing the area of the nail tip
   B. decreasing the area of the nail tip
   C. increasing the length of the nail
   D. decreasing the length of the nail

Test-Taking Tip

Check the Answer Number For each question, double check that you are filling in the correct answer bubble for the question number you are working on.

4. A 15-g block of aluminum has a volume of 5.5 cm³. What is the density?
   A. 2.7 g/cm³
   B. 82.5 g/cm³
   C. 0.37 g/cm³
   D. 2.7 cm³/g

Use the illustration below to answer questions 5 and 6.

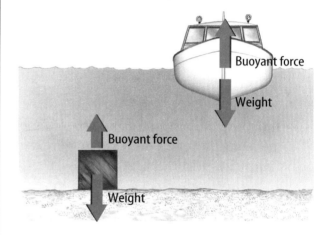

5. Assume the boat and cube have the same mass. Which of these is correct?
   A. The boat displaces less water than the cube.
   B. The densities of the boat and cube are equal.
   C. The density of the boat is less than the density of the cube.
   D. The density of the boat is greater than the density of the water.

6. Which of the following would make the cube more likely to float?
   A. increasing its volume
   B. increasing its density
   C. increasing its weight
   D. decreasing its volume

7. Which of the following instruments is used to measure atmospheric pressure?
   A. altimeter
   B. hygrometer
   C. barometer
   D. anemometer

Standardized Test Practice

Part 2 Short Response/Grid In

Record your answers on the answer sheet provided by your teacher or on a sheet of paper.

8. Explain why wearing snowshoes makes it easier to walk over deep, soft snow.

9. Why is a gas, such as air condidered to be a fluid?

10. Explain why the pressure exerted by the atmosphere does not crush your body.

Use the illustration below to answer questions 11 and 12.

11. How does the buoyant force on the boat change if the boat is loaded so that it floats lower in the water?

12. What changes in the properties of this boat could cause it to sink?

13. People sometimes prepare for a coming hurricane by nailing boards over the outside of windows. How can high winds damage windows?

14. What factors influence the amount of lift on an airplane wing?

15. The pressure inside the fluid of a hydraulic system is increased by 1,000 Pa. What is the increase in the force exerted on a piston that has an area of 0.05 m²?

Part 3 Open Ended

Record your answers on a sheet of paper.

16. Explain why the interior of an airplane is pressurized when flying at high altitude. If a hole were punctured in an exterior wall, what would happen to the air pressure inside the plane?

17. In an experiment, you design small boats using aluminum foil. You add pennies to each boat until it sinks. Sketch several different shapes you might create. Which will likely hold the most pennies? Why?

18. You squeeze a round, air-filled balloon slightly, changing its shape. Describe how the pressure inside the balloon changes.

Use the illustration below to answer questions 19 and 20.

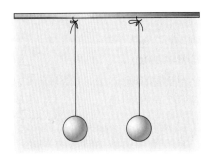

19. Describe the motion of the table tennis balls when air is blown between the two balls. Explain.

20. Describe the motion of the table tennis balls if air blown to the left of the ball on the left and to the right of the ball on the right at the same time. Explain.

21. Compare the pressure you exert on a floor when you are standing on the floor and when you are lying on the floor.

22. In order to drink a milk shake through a straw, how must the air pressure inside the straw change, compared with the air pressure outside the straw?

chapter 4

Work and Simple Machines

The BIG Idea
A machine makes doing a job easier.

SECTION 1
Work and Power
Main Idea Work is done when a force causes an object to move in the same direction as the force.

SECTION 2
Using Machines
Main Idea A machine can change the force needed to do a job.

SECTION 3
Simple Machines
Main Idea There are six types of simple machines.

Heavy Lifting

It took the ancient Egyptians more than 100 years to build the pyramids without machines like these. But now, even tall skyscrapers can be built in a few years. Complex or simple, machines have the same purpose. They make doing a job easier.

Science Journal Describe three machines you used today, and how they made doing a task easier.

Start-Up Activities

Compare Forces

Two of the world's greatest structures were built using different tools. The Great Pyramid at Giza in Egypt was built nearly 5,000 years ago using blocks of limestone moved into place by hand with ramps and levers. In comparison, the Sears Tower in Chicago was built in 1973 using tons of steel that were hoisted into place by gasoline-powered cranes. How do machines such as ramps, levers, and cranes change the forces needed to do a job?

1. Place a ruler on an eraser. Place a book on one end of the ruler.
2. Using one finger, push down on the free end of the ruler to lift the book.
3. Repeat the experiment, placing the eraser in various positions beneath the ruler. Observe how much force is needed in each instance to lift the book.
4. **Think Critically** In your Science Journal, describe your observations. How did changing the distance between the book and the eraser affect the force needed to lift the book?

Simple Machines Many of the devices that you use every day are simple machines. Make the following Foldable to help you understand the characteristics of simple machines.

STEP 1 Draw a mark at the midpoint of a sheet of paper along the side edge. Then fold the top and bottom edges in to touch the midpoint.

STEP 2 Fold in half from side to side.

STEP 3 Turn the paper vertically. Open and cut along the inside fold lines to form four tabs.

STEP 4 Label the tabs *Inclined Plane, Lever, Wheel and Axle,* and *Pulley.*

Read for Main Ideas As you read the chapter, list the characteristics of inclined planes, levers, wheels and axles, and pulleys under the appropriate tab.

Preview this chapter's content and activities at bookm.msscience.com

Get Ready to Read

Questions and Answers

① Learn It! Knowing how to find answers to questions will help you on reviews and tests. Some answers can be found in the textbook, while other answers require you to go beyond the textbook. These answers might be based on knowledge you already have or things you have experienced.

② Practice It! Read the excerpt below. Answer the following questions and then discuss them with a partner.

> Did you use a machine today? When you think of a machine, you might think of a device, such as a car, with many moving parts powered by an engine or an electric motor. But if you used a pair of scissors or a broom, or cut your food with a knife, you used a machine. A machine is simply a device that makes doing work easier. Even a sloping surface can be a machine.
>
> —from page 104

- Describe how using a broom makes cleaning a floor easier.
- How is pushing a box up a smooth ramp easier than lifting the box upward?
- Why does a screwdriver make it easier to tighten a screw?

③ Apply It! Look at some questions in the text. Which questions can be answered directly from the text? Which require you to go beyond the text?

Target Your Reading

As you read, keep track of questions you answer in the chapter. This will help you remember what you read.

Use this to focus on the main ideas as you read the chapter.

1 Before you read the chapter, respond to the statements below on your worksheet or on a numbered sheet of paper.
- Write an **A** if you **agree** with the statement.
- Write a **D** if you **disagree** with the statement.

2 After you read the chapter, look back to this page to see if you've changed your mind about any of the statements.
- If any of your answers changed, explain why.
- Change any false statements into true statements.
- Use your revised statements as a study guide.

Science Online
Print out a worksheet of this page at bookm.msscience.com

Before You Read A or D	Statement	After You Read A or D
	1 Friction is caused by atoms or molecules of one object bonding to atoms or molecules in another object.	
	2 Power measures how fast work is done.	
	3 The fulcrum of a lever is always between the input force and the output force.	
	4 Efficiency is the ratio of output work to input work.	
	5 When you do work on an object, you transfer energy to the object.	
	6 A car is a combination of simple machines.	
	7 A wedge and a screw are both types of inclined planes.	
	8 Work is done anytime a force is applied.	
	9 Mechanical advantage can never be less than 1.	

M ◆ 98 B

section 1
Work and Power

as you read

What You'll Learn
- **Recognize** when work is done.
- **Calculate** how much work is done.
- **Explain** the relation between work and power.

Why It's Important
If you understand work, you can make your work easier.

Review Vocabulary
force: a push or a pull

New Vocabulary
- work
- power

What is work?

What does the term *work* mean to you? You might think of household chores; a job at an office, a factory, a farm; or the homework you do after school. In science, the definition of work is more specific. **Work** is done when a force causes an object to move in the same direction that the force is applied.

Can you think of a way in which you did work today? Maybe it would help to know that you do work when you lift your books, turn a doorknob, raise window blinds, or write with a pen or pencil. You also do work when you walk up a flight of stairs or open and close your school locker. In what other ways do you do work every day?

Work and Motion Your teacher has asked you to move a box of books to the back of the classroom. Try as you might, though, you just can't budge the box because it is too heavy. Although you exerted a force on the box and you feel tired from it, you have not done any work. In order for you to do work, two things must occur. First, you must apply a force to an object. Second, the object must move in the same direction as your applied force. You do work on an object only when the object moves as a result of the force you exert. The girl in **Figure 1** might think she is working by holding the bags of groceries. However, if she is not moving, she is not doing any work because she is not causing something to move.

Reading Check *To do work, how must a force make an object move?*

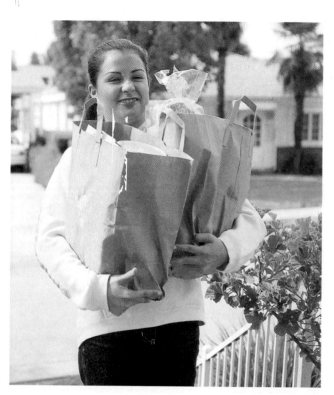

Figure 1 This girl is holding bags of groceries, yet she isn't doing any work. **Explain** *what must happen for work to be done.*

Figure 2 To do work, an object must move in the direction a force is applied.

The boy's arms do work when they exert an upward force on the basket and the basket moves upward.

The boy's arms still exert an upward force on the basket. But when the boy walks forward, no work is done by his arms.

Applying Force and Doing Work Picture yourself lifting the basket of clothes in **Figure 2.** You can feel your arms exerting a force upward as you lift the basket, and the basket moves upward in the direction of the force your arms applied. Therefore, your arms have done work. Now, suppose you carry the basket forward. You can still feel your arms applying an upward force on the basket to keep it from falling, but now the basket is moving forward instead of upward. Because the direction of motion is not in the same direction of the force applied by your arms, no work is done by your arms.

Force in Two Directions Sometimes only part of the force you exert moves an object. Think about what happens when you push a lawn mower. You push at an angle to the ground as shown in **Figure 3.** Part of the force is to the right and part of the force is downward. Only the part of the force that is in the same direction as the motion of the mower—to the right—does work.

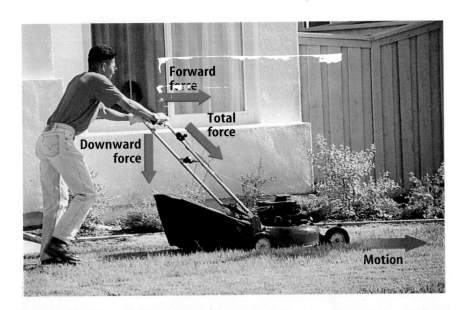

Figure 3 When you exert a force at an angle, only part of your force does work—the part that is in the same direction as the motion of the object.

James Prescott Joule This English physicist experimentally verified the law of conservation of energy. He showed that various forms of energy—mechanical, electrical, and thermal—are essentially the same and can be converted one into another. The SI unit of energy and work, the joule, is named after him. Research the work of Joule and write what you learn in your Science Journal.

Calculating Work

Work is done when a force makes an object move. More work is done when the force is increased or the object is moved a greater distance. Work can be calculated using the work equation below. In SI units, the unit for work is the joule, named for the nineteenth-century scientist James Prescott Joule.

Work Equation

work (in joules) = **force** (in newtons) × **distance** (in meters)

$$W = Fd$$

Work and Distance Suppose you give a book a push and it slides across a table. To calculate the work you did, the distance in the above equation is not the distance the book moved. The distance in the work equation is the distance an object moves while the force is being applied. So the distance in the work equation is the distance the book moved while you were pushing.

Applying Math — Solve a One-Step Equation

CALCULATING WORK A painter lifts a can of paint that weighs 40 N a distance of 2 m. How much work does she do? *Hint: to lift a can weighing 40 N, the painter must exert a force of 40 N.*

Solution

1. *This is what you know:*
   - force: $F = 40$ N
   - distance: $d = 2$ m

2. *This is what you need to find out:*
   work: $W = ?$ J

3. *This is the procedure you need to use:*
   Substitute the known values $F = 40$ N and $d = 2$ m into the work equation:
   $W = Fd = (40\text{ N})(2\text{ m}) = 80$ N·m $= 80$ J

4. *Check your answer:*
   Check your answer by dividing the work you calculated by the distance given in the problem. The result should be the force given in the problem.

Practice Problems

1. As you push a lawn mower, the horizontal force is 300 N. If you push the mower a distance of 500 m, how much work do you do?

2. A librarian lifts a box of books that weighs 93 N a distance of 1.5 m. How much work does he do?

For more practice, visit
bookm.msscience.com/
math_practice

What is power?

What does it mean to be powerful? Imagine two weightlifters lifting the same amount of weight the same vertical distance. They both do the same amount of work. However, the amount of power they use depends on how long it took to do the work. **Power** is how quickly work is done. The weightlifter who lifted the weight in less time is more powerful.

Calculating Power Power can be calculated by dividing the amount of work done by the time needed to do the work.

Power Equation

$$\text{power (in watts)} = \frac{\text{work (in joules)}}{\text{time (in seconds)}}$$

$$P = \frac{W}{t}$$

In SI units, the unit of power is the watt, in honor of James Watt, a nineteenth-century British scientist who invented a practical version of the steam engine.

Mini LAB

Work and Power

Procedure
1. Weigh yourself on a **scale**.
2. Multiply your weight in pounds by 4.45 to convert your weight to newtons.
3. Measure the vertical height of a **stairway**. **WARNING:** *Make sure the stairway is clear of all objects.*
4. Time yourself walking slowly and quickly up the stairway.

Analysis
Calculate and compare the work and power in each case.

Try at Home

Applying Math — Solve a One-Step Equation

CALCULATING POWER You do 200 J of work in 12 s. How much power did you use?

Solution

1 *This is what you know:*
- work: $W = 200$ J
- time: $t = 12$ s

2 *This is what you need to find out:*
- power: $P = ?$ watts

3 *This is the procedure you need to use:*

Substitute the known values $W = 200$ J and $t = 12$ s into the power equation:

$$P = \frac{W}{t} = \frac{200 \text{ J}}{12 \text{ s}} = 17 \text{ watts}$$

4 *Check your answer:*

Check your answer by multiplying the power you calculated by the time given in the problem. The result should be the work given in the problem.

Practice Problems

1. In the course of a short race, a car does 50,000 J of work in 7 s. What is the power of the car during the race?
2. A teacher does 140 J of work in 20 s. How much power did he use?

For more practice, visit bookm.msscience.com/math_practice

SECTION 1 Work and Power

Topic: James Watt
Visit bookm.msscience.com for Web links to information about James Watt and his steam engine.

Activity Draw a diagram showing how his steam engine worked.

Work and Energy If you push a chair and make it move, you do work on the chair and change its energy. Recall that when something is moving it has energy of motion, or kinetic energy. By making the chair move, you increase its kinetic energy.

You also change the energy of an object when you do work and lift it higher. An object has potential energy that increases when it is higher above Earth's surface. By lifting an object, you do work and increase its potential energy.

Power and Energy When you do work on an object you increase the energy of the object. Because energy can never be created or destroyed, if the object gains energy then you must lose energy. When you do work on an object you transfer energy to the object, and your energy decreases. The amount of work done is the amount of energy transferred. So power is also equal to the amount of energy transferred in a certain amount of time.

Sometimes energy can be transferred even when no work is done, such as when heat flows from a warm to a cold object. In fact, there are many ways energy can be transferred even if no work is done. Power is always the rate at which energy is transferred, or the amount of energy transferred divided by the time needed.

section 1 review

Summary

What is work?
- Work is done when a force causes an object to move in the same direction that the force is applied.
- If the movement caused by a force is at an angle to the direction the force is applied, only the part of the force in the direction of motion does work.
- Work can be calculated by multiplying the force applied by the distance:

$$W = Fd$$

- The distance in the work equation is the distance an object moves while the force is being applied.

What is power?
- Power is how quickly work is done. Something is more powerful if it can do a given amount of work in less time.
- Power can be calculated by dividing the work done by the time needed to do the work:

$$P = \frac{W}{t}$$

Self Check

1. **Describe** a situation in which work is done on an object.
2. **Evaluate** which of the following situations involves more power: 200 J of work done in 20 s or 50 J of work done in 4 s? Explain your answer.
3. **Determine** two ways power can be increased.
4. **Calculate** how much power, in watts, is needed to cut a lawn in 50 min if the work involved is 100,000 J.
5. **Think Critically** Suppose you are pulling a wagon with the handle at an angle. How can you make your task easier?

Applying Math

6. **Calculate Work** How much work was done to lift a 1,000-kg block to the top of the Great Pyramid, 146 m above ground?
7. **Calculate Work Done by an Engine** An engine is used to lift a beam weighing 9,800 N up to 145 m. How much work must the engine do to lift this beam? How much work must be done to lift it 290 m?

Building the Pyramids

Imagine moving 2.3 million blocks of limestone, each weighing more than 1,000 kg. That is exactly what the builders of the Great Pyramid at Giza did. Although no one knows for sure exactly how they did it, they probably pulled the blocks most of the way.

◉ Real-World Question

How is the force needed to lift a block related to the distance it travels?

Goals
■ **Compare** the force needed to lift a block with the force needed to pull it up a ramp.

Materials
wood block
tape
spring scale
ruler
thin notebooks
meterstick
several books

Safety Precautions

Work Done Using Different Ramps		
Distance (cm)	Force (N)	Work (J)
Do not write in this book.		

◉ Procedure

1. Stack several books together on a tabletop to model a half-completed pyramid. Measure the height of the books in centimeters. Record the height on the first row of the data table under *Distance*.

2. Use the wood block as a model for a block of stone. Use tape to attach the block to the spring scale.

3. Place the block on the table and lift it straight up the side of the stack of books until the top of the block is even with the top of the books. Record the force shown on the scale in the data table under *Force*.

4. **Arrange** a notebook so that one end is on the stack of books and the other end is on the table. Measure the length of the notebook and record this length as distance in the second row of the data table under *Distance*.

5. **Measure** the force needed to pull the block up the ramp. Record the force in the data table.

6. Repeat steps 4 and 5 using a longer notebook to make the ramp longer.

7. **Calculate** the work done in each row of the data table.

◉ Conclude and Apply

1. **Evaluate** how much work you did in each instance.

2. **Determine** what happened to the force needed as the length of the ramp increased.

3. **Infer** How could the builders of the pyramids have designed their task to use less force than they would lifting the blocks straight up? Draw a diagram to support your answer.

*C*ommunicating Your Data

Add your data to that found by other groups. **For more help, refer to the** Science Skill Handbook.

section 2
Using Machines

as you read

What You'll Learn
- **Explain** how a machine makes work easier.
- **Calculate** the mechanical advantages and efficiency of a machine.
- **Explain** how friction reduces efficiency.

Why It's Important
Machines can't change the amount of work you need to do, but they can make doing work easier.

Review Vocabulary
friction: force that opposes motion between two touching surfaces

New Vocabulary
- input force
- output force
- mechanical advantage
- efficiency

What is a machine?

Did you use a machine today? When you think of a machine you might think of a device, such as a car, with many moving parts powered by an engine or an electric motor. But if you used a pair of scissors or a broom, or cut your food with a knife, you used a machine. A machine is simply a device that makes doing work easier. Even a sloping surface can be a machine.

Mechanical Advantage

Even though machines make work easier, they don't decrease the amount of work you need to do. Instead, a machine changes the way in which you do work. When you use a machine, you exert a force over some distance. For example, you exert a force to move a rake or lift the handles of a wheelbarrow. The force that you apply on a machine is the **input force**. The work you do on the machine is equal to the input force times the distance over which your force is applied. The work that you do on the machine is the input work.

The machine also does work by exerting a force to move an object over some distance. A rake, for example, exerts a force to move leaves. Sometimes this force is called the resistance force because the machine is trying to overcome some resistance. The force that the machine applies is the **output force**. The work that the machine does is the output work. **Figure 4** shows how a machine transforms input work to output work.

When you use a machine, the output work can never be greater than the input work. So what is the advantage of using a machine? A machine makes work easier by changing the amount of force you need to exert, the distance over which the force is exerted, or the direction in which you exert your force.

Figure 4 No matter what type of machine is used, the output work is never greater than the input work.

Changing Force Some machines make doing a job easier by reducing the force you have to apply to do the job. For this type of machine the output force is greater than the input force. How much larger the output force is compared to the input force is the **mechanical advantage** of the machine. The mechanical advantage of a machine is the ratio of the output force to the input force and can be calculated from this equation:

Mechanical Advantage Equation

$$\text{mechanical advantage} = \frac{\text{output force (in newtons)}}{\text{input force (in newtons)}}$$

$$MA = \frac{F_{out}}{F_{in}}$$

Mechanical advantage does not have any units, because it is the ratio of two numbers with the same units.

Topic: Historical Tools
Visit bookm.msscience.com for Web links to information about early types of tools and how they took advantage of simple machines.

Activity Write a paragraph describing how simple machines were used to design early tools.

Applying Math — Solve a One-Step Equation

CALCULATING MECHANICAL ADVANTAGE To pry the lid off a paint can, you apply a force of 50 N to the handle of the screwdriver. What is the mechanical advantage of the screwdriver if it applies a force of 500 N to the lid?

Solution

1 This is what you know:
- input force: F_{in} = 50 N
- output force: F_{out} = 500 N

2 This is what you need to find out: mechanical advantage: MA = ?

3 This is the procedure you need to use: Substitute the known values F_{in} = 50 N and F_{out} = 500 N into the mechanical advantage equation:

$$MA = \frac{F_{out}}{F_{in}} = \frac{500 \text{ N}}{50 \text{ N}} = 10$$

4 Check your answer: Check your answer by multiplying the mechanical advantage you calculated by the input force given in the problem. The result should be the output force given in the problem.

Practice Problems

1. To open a bottle, you apply a force of 50 N to the bottle opener. The bottle opener applies a force of 775 N to the bottle cap. What is the mechanical advantage of the bottle opener?

2. To crack a pecan, you apply a force of 50 N to the nutcracker. The nutcracker applies a force of 750 N to the pecan. What is the mechanical advantage of the nutcracker?

For more practice, visit bookm.msscience.com/math_practice

SECTION 2 Using Machines

Figure 5 Changing the direction or the distance that a force is applied can make a task easier.

Sometimes it is easier to exert your force in a certain direction. This boy would rather pull down on the rope to lift the flag than to climb to the top of the pole and pull up.

When you rake leaves, you move your hands a short distance, but the end of the rake moves over a longer distance.

Changing Distance Some machines allow you to exert your force over a shorter distance. In these machines, the output force is less than the input force. The rake in **Figure 5** is this type of machine. You move your hands a small distance at the top of the handle, but the bottom of the rake moves a greater distance as it moves the leaves. The mechanical advantage of this type of machine is less than one because the output force is less than the input force.

Changing Direction Sometimes it is easier to apply a force in a certain direction. For example, it is easier to pull down on the rope in **Figure 5** than to pull up on it. Some machines enable you to change the direction of the input force. In these machines neither the force nor the distance is changed. The mechanical advantage of this type of machine is equal to one because the output force is equal to the input force. The three ways machines make doing work easier are summarized in **Figure 6**.

Figure 6 Machines are useful because they can increase force, increase distance, or change the direction in which a force is applied.

Efficiency

A machine can't make the output work greater than the input work. In fact, for a real machine, the output work is always less than the input work. In a real machine, there is friction as parts of the machine move. Friction converts some of the input work into heat, so that the output work is reduced. The **efficiency** of a machine is the ratio of the output work to the input work, and can be calculated from this equation:

Efficiency Equation

$$\text{efficiency (in percent)} = \frac{\text{output work (in joules)}}{\text{input work (in joules)}} \times 100\%$$

$$\text{eff} = \frac{W_{\text{out}}}{W_{\text{in}}} \times 100\%$$

If the amount of friction in the machine is reduced, the efficiency of the machine increases.

Body Temperature
Chemical reactions that enable your muscles to move also produce heat that helps maintain your body temperature. When you shiver, rapid contraction and relaxation of muscle fibers produces a large amount of heat that helps raise your body temperature. This causes the efficiency of your muscles to decrease as more energy is converted into heat.

Applying Math — Solve a One-Step Equation

CALCULATING EFFICIENCY Using a pulley system, a crew does 7,500 J of work to load a box that requires 4,500 J of work. What is the efficiency of the pulley system?

Solution

1 *This is what you know:*
- input work: $W_{\text{in}} = 7{,}500$ J
- output work: $W_{\text{out}} = 4{,}500$ J

2 *This is what you need to find out:* efficiency: $\text{eff} = ?\ \%$

3 *This is the procedure you need to use:* Substitute the known values $W_{\text{in}} = 7{,}500$ J and $W_{\text{out}} = 4{,}500$ J into the efficiency equation:

$$\text{eff} = \frac{W_{\text{out}}}{W_{\text{in}}} = \frac{4{,}500 \text{ J}}{7{,}500 \text{ J}} \times 100\% = 60\%$$

4 *Check your answer:* Check your answer by dividing the efficiency by 100% and then multiplying your answer times the work input. The product should be the work output given in the problem.

Practice Problems

1. You do 100 J of work in pulling out a nail with a claw hammer. If the hammer does 70 J of work, what is the hammer's efficiency?

2. You do 150 J of work pushing a box up a ramp. If the ramp does 105 J of work, what is the efficiency of the ramp?

For more practice, visit bookm.msscience.com/math_practice

Figure 7 Lubrication can reduce the friction between two surfaces. Two surfaces in contact can stick together where the high spots on each surface come in contact. Adding oil or another lubricant separates the surface so that fewer high spots make contact.

Friction To help understand friction, imagine pushing a heavy box up a ramp. As the box begins to move, the bottom surface of the box slides across the top surface of the ramp. Neither surface is perfectly smooth—each has high spots and low spots, as shown in **Figure 7**.

As the two surfaces slide past each other, high spots on the two surfaces come in contact. At these contact points, shown in **Figure 7,** atoms and molecules can bond together. This makes the contact points stick together. The attractive forces between all the bonds in the contact points added together is the frictional force that tries to keep the two surfaces from sliding past each other.

To keep the box moving, a force must be applied to break the bonds between the contact points. Even after these bonds are broken and the box moves, new bonds form as different parts of the two surfaces come into contact.

Friction and Efficiency One way to reduce friction between two surfaces is to add oil. **Figure 7** shows how oil fills the gaps between the surfaces, and keeps many of the high spots from making contact. Because there are fewer contact points between the surfaces, the force of friction is reduced. More of the input work then is converted to output work by the machine.

section 2 review

Summary

What is a machine?
- A machine is a device that makes doing work easier.
- A machine can make doing work easier by reducing the force exerted, changing the distance over which the force is exerted, or changing the direction of the force.
- The output work done by a machine can never be greater than the input work done on the machine.

Mechanical Advantage and Efficiency
- The mechanical advantage of a machine is the number of times the machine increases the input force:
$$MA = \frac{F_{out}}{F_{in}}$$
- The efficiency of a machine is the ratio of the output work to the input work:
$$eff = \frac{W_{out}}{W_{in}} \times 100\%$$

Self Check

1. **Identify** three specific situations in which machines make work easier.
2. **Infer** why the output force exerted by a rake must be less than the input force.
3. **Explain** how the efficiency of an ideal machine compares with the efficiency of a real machine.
4. **Explain** how friction reduces the efficiency of machines.
5. **Think Critically** Can a machine be useful even if its mechanical advantage is less than one? Explain and give an example.

Applying Math

6. **Calculate Efficiency** Find the efficiency of a machine if the input work is 150 J and the output work is 90 J.
7. **Calculate Mechanical Advantage** To lift a crate, a pulley system exerts a force of 2,750 N. Find the mechanical advantage of the pulley system if the input force is 250 N.

section 3 Simple Machines

What is a simple machine?

What do you think of when you hear the word *machine*? Many people think of machines as complicated devices such as cars, elevators, or computers. However, some machines are as simple as a hammer, shovel, or ramp. A **simple machine** is a machine that does work with only one movement. The six simple machines are the inclined plane, lever, wheel and axle, screw, wedge, and pulley. A machine made up of a combination of simple machines is called a **compound machine.** A can opener is a compound machine. The bicycle in **Figure 8** is a familiar example of another compound machine.

Inclined Plane

Ramps might have enabled the ancient Egyptians to build their pyramids. To move limestone blocks weighing more than 1,000 kg each, archaeologists hypothesize that the Egyptians built enormous ramps. A ramp is a simple machine known as an inclined plane. An **inclined plane** is a flat, sloped surface. Less force is needed to move an object from one height to another using an inclined plane than is needed to lift the object. As the inclined plane becomes longer, the force needed to move the object becomes smaller.

as you read

What You'll Learn
- **Distinguish** among the different simple machines.
- **Describe** how to find the mechanical advantage of each simple machine.

Why It's Important
All machines, no matter how complicated, are made of simple machines.

Review Vocabulary
compound: made of separate pieces or parts

New Vocabulary
- simple machine
- compound machine
- inclined plane
- wedge
- screw
- lever
- wheel and axle
- pulley

Figure 8 Devices that use combinations of simple machines, such as this bicycle, are called compound machines.

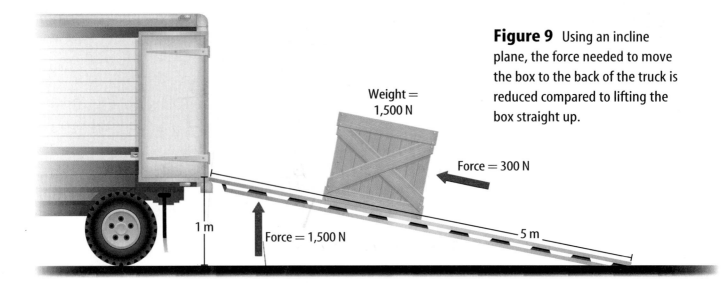

Figure 9 Using an incline plane, the force needed to move the box to the back of the truck is reduced compared to lifting the box straight up.

Using Inclined Planes Imagine having to lift a box weighing 1,500 N to the back of a truck that is 1 m off the ground. You would have to exert a force of 1,500 N, the weight of the box, over a distance of 1 m, which equals 1,500 J of work. Now suppose that instead you use a 5-m-long ramp, as shown in **Figure 9**. The amount of work you need to do does not change. You still need to do 1,500 J of work. However, the distance over which you exert your force becomes 5 m. You can calculate the force you need to exert by dividing both sides of the equation for work by distance.

$$\text{Force} = \frac{\text{work}}{\text{distance}}$$

If you do 1,500 J of work by exerting a force over 5 m, the force is only 300 N. Because you exert the input force over a distance that is five times as long, you can exert a force that is five times less.

The mechanical advantage of an inclined plane is the length of the inclined plane divided by its height. In this example, the ramp has a mechanical advantage of 5.

Wedge An inclined plane that moves is called a **wedge.** A wedge can have one or two sloping sides. The knife shown in **Figure 10** is an example of a wedge. An axe and certain types of doorstops are also wedges. Just as for an inclined plane, the mechanical advantage of a wedge increases as it becomes longer and thinner.

Figure 10 This chef's knife is a wedge that slices through food.

Figure 11 Wedge-shaped teeth help tear food.

Your front teeth help tear an apple apart.

The wedge-shaped teeth of this *Tyrannosaurus rex* show that it was a carnivore.

Wedges in Your Body You have wedges in your body. The bite marks on the apple in **Figure 11** show how your front teeth are wedge shaped. A wedge changes the direction of the applied effort force. As you push your front teeth into the apple, the downward effort force is changed by your teeth into a sideways force that pushes the skin of the apple apart.

The teeth of meat eaters, or carnivores, are more wedge shaped than the teeth of plant eaters, or herbivores. The teeth of carnivores are used to cut and rip meat, while herbivores' teeth are used for grinding plant material. By examining the teeth of ancient animals, such as the dinosaur in **Figure 11,** scientists can determine what the animal ate when it was living.

The Screw Another form of the inclined plane is a screw. A **screw** is an inclined plane wrapped around a cylinder or post. The inclined plane on a screw forms the screw threads. Just like a wedge changes the direction of the effort force applied to it, a screw also changes the direction of the applied force. When you turn a screw, the force applied is changed by the threads to a force that pulls the screw into the material. Friction between the threads and the material holds the screw tightly in place. The mechanical advantage of the screw is the length of the inclined plane wrapped around the screw divided by the length of the screw. The more tightly wrapped the threads are, the easier it is to turn the screw. Examples of screws are shown in **Figure 12.**

Figure 12 The thread around a screw is an inclined plane. Many familiar devices use screws to make work easier.

 How are screws related to the inclined plane?

Figure 13 The mechanical advantage of a lever changes as the position of the fulcrum changes. The mechanical advantage increases as the fulcrum is moved closer to the output force.

Figure 14 A faucet handle is a wheel and axle. A wheel and axle is similar to a circular lever. The center is the fulcrum, and the wheel and axle turn around it. **Explain** *how you can increase the mechanical advantage of a wheel and axle.*

Lever

You step up to the plate. The pitcher throws the ball and you swing your lever to hit the ball? That's right! A baseball bat is a type of simple machine called a lever. A **lever** is any rigid rod or plank that pivots, or rotates, about a point. The point about which the lever pivots is called a fulcrum.

The mechanical advantage of a lever is found by dividing the distance from the fulcrum to the input force by the distance from the fulcrum to the output force, as shown in **Figure 13**. When the fulcrum is closer to the output force than the input force, the mechanical advantage is greater than one.

Levers are divided into three classes according to the position of the fulcrum with respect to the input force and output force. **Figure 15** shows examples of three classes of levers.

Wheel and Axle

Do you think you could turn a doorknob easily if it were a narrow rod the size of a pencil? It might be possible, but it would be difficult. A doorknob makes it easier for you to open a door because it is a simple machine called a wheel and axle. A **wheel and axle** consists of two circular objects of different sizes that are attached in such a way that they rotate together. As you can see in **Figure 14,** the larger object is the wheel and the smaller object is the axle.

The mechanical advantage of a wheel and axle is usually greater than one. It is found by dividing the radius of the wheel by the radius of the axle. For example, if the radius of the wheel is 12 cm and the radius of the axle is 4 cm, the mechanical advantage is 3.

112 ◆ **M** CHAPTER 4 Work and Simple Machines

NATIONAL GEOGRAPHIC VISUALIZING LEVERS

Figure 15

Levers are among the simplest of machines, and you probably use them often in everyday life without even realizing it. A lever is a bar that pivots around a fixed point called a fulcrum. As shown here, there are three types of levers—first class, second class, and third class. They differ in where two forces—an input force and an output force—are located in relation to the fulcrum.

 Fulcrum
 Input force
Output force

In a first-class lever, the fulcrum is between the input force and the output force. First-class levers, such as scissors and pliers, multiply force or distance depending on where the fulcrum is placed. They always change the direction of the input force, too.

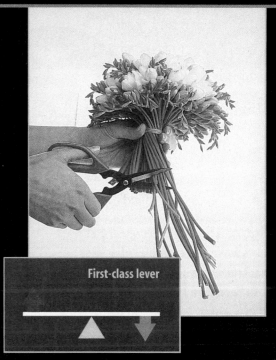

First-class lever

In a second-class lever, such as a wheelbarrow, the output force is between the input force and the fulcrum. Second-class levers always multiply the input force but don't change its direction.

Second-class lever

Third-class lever

In a third-class lever, such as a baseball bat, the input force is between the output force and the fulcrum. For a third-class lever, the output force is less than the input force, but is in the same direction.

SECTION 3 Simple Machines M ◆ 113

Figure 16 The waterwheel and Ferris wheel are examples of devices that rely on a wheel and axle. **Compare and contrast** waterwheels and Ferris wheels in terms of wheels and axles.

Observing Pulleys

Procedure
1. Obtain two **broomsticks**. Tie a 3-m-long **rope** to the middle of one stick. Wrap the rope around both sticks four times.
2. Have two students pull the broomsticks apart while a third pulls on the rope.
3. Repeat with two wraps of rope.

Analysis
1. Compare the results.
2. Predict whether it will be easier to pull the broomsticks together with ten wraps of rope.

Using Wheels and Axles In some devices, the input force is used to turn the wheel and the output force is exerted by the axle. Because the wheel is larger than the axle, the mechanical advantage is greater than one. So the output force is greater than the input force. A doorknob, a steering wheel, and a screwdriver are examples of this type of wheel and axle.

In other devices, the input force is applied to turn the axle and the output force is exerted by the wheel. Then the mechanical advantage is less than one and the output force is less than the input force. A fan and a ferris wheel are examples of this type of wheel and axle. **Figure 16** shows an example of each type of wheel and axle.

Pulley

To raise a sail, a sailor pulls down on a rope. The rope uses a simple machine called a pulley to change the direction of the force needed. A **pulley** consists of a grooved wheel with a rope or cable wrapped over it.

Fixed Pulleys Some pulleys, such as the one on a sail, a window blind, or a flagpole, are attached to a structure above your head. When you pull down on the rope, you pull something up. This type of pulley, called a fixed pulley, does not change the force you exert or the distance over which you exert it. Instead, it changes the direction in which you exert your force, as shown in **Figure 17**. The mechanical advantage of a fixed pulley is 1.

Reading Check *How does a fixed pulley affect the input force?*

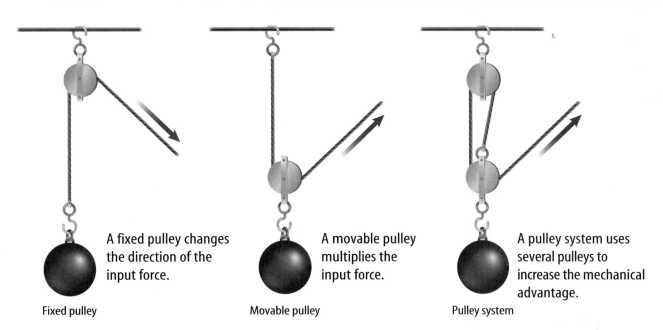

Figure 17 Pulleys can change force and direction.

Movable Pulleys Another way to use a pulley is to attach it to the object you are lifting, as shown in **Figure 17**. This type of pulley, called a movable pulley, allows you to exert a smaller force to lift the object. The mechanical advantage of a movable pulley is always 2.

More often you will see combinations of fixed and movable pulleys. Such a combination is called a pulley system. The mechanical advantage of a pulley system is equal to the number of sections of rope pulling up on the object. For the pulley system shown in **Figure 17** the mechanical advantage is 3.

section 3 review

Summary

Simple and Compound Machines
- A simple machine is a machine that does work with only one movement.
- A compound machine is made from a combination of simple machines.

Types of Simple Machines
- An inclined plane is a flat, sloped surface.
- A wedge is an inclined plane that moves.
- A screw is an inclined plane that is wrapped around a cylinder or post.
- A lever is a rigid rod that pivots around a fixed point called the fulcrum.
- A wheel and axle consists of two circular objects of different sizes that rotate together.
- A pulley is a grooved wheel with a rope or cable wrapped over it.

Self Check

1. **Determine** how the mechanical advantage of a ramp changes as the ramp becomes longer.
2. **Explain** how a wedge changes an input force.
3. **Identify** the class of lever for which the fulcrum is between the input force and the output force.
4. **Explain** how the mechanical advantage of a wheel and axle change as the size of the wheel increases.
5. **Think Critically** How are a lever and a wheel and axle similar?

Applying Math

6. **Calculate Length** The Great Pyramid is 146 m high. How long is a ramp from the top of the pyramid to the ground that has a mechanical advantage of 4?
7. **Calculate Force** Find the output force exerted by a moveable pulley if the input force is 50 N.

LAB Design Your Own

Pulley Power

Real-World Question

Imagine how long it might have taken to build the Sears Tower in Chicago without the aid of a pulley system attached to a crane. Hoisting the 1-ton I beams to a maximum height of 110 stories required large lifting forces and precise control of the beam's movement.

Construction workers also use smaller pulleys that are not attached to cranes to lift supplies to where they are needed. Pulleys are not limited to construction sites. They also are used to lift automobile engines out of cars, to help load and unload heavy objects on ships, and to lift heavy appliances and furniture. How can you use a pulley system to reduce the force needed to lift a load?

Goals
- **Design** a pulley system.
- **Measure** the mechanical advantage and efficiency of the pulley system.

Possible Materials
single- and multiple-pulley systems
nylon rope
steel bar to support the pulley system
meterstick
*metric tape measure
variety of weights to test pulleys
force spring scale
brick
*heavy book
balance
*scale
*Alternate materials

Safety Precautions

WARNING: *The brick could be dangerous if it falls. Keep your hands and feet clear of it.*

Form a Hypothesis

Write a hypothesis about how pulleys can be combined to make a system of pulleys to lift a heavy load, such as a brick. Consider the efficiency of your system.

Test Your Hypothesis

Make a Plan

1. Decide how you are going to support your pulley system. What materials will you use?
2. How will you measure the effort force and the resistance force? How will you determine the mechanical advantage? How will you measure efficiency?
3. **Experiment** by lifting small weights with a single pulley, double pulley, and so on. How efficient are the pulleys? In what ways can you increase the efficiency of your setup?

Using Scientific Methods

4. Use the results of step 3 to design a pulley system to lift the brick. Draw a diagram of your design. Label the different parts of the pulley system and use arrows to indicate the direction of movement for each section of rope.

Follow Your Plan

1. Make sure your teacher approves your plan before you start.
2. Assemble the pulley system you designed. You might want to test it with a smaller weight before attaching the brick.
3. **Measure** the force needed to lift the brick. How much rope must you pull to raise the brick 10 cm?

Analyze Your Data

1. **Calculate** the ideal mechanical advantage of your design.
2. **Calculate** the actual mechanical advantage of the pulley system you built.
3. **Calculate** the efficiency of your pulley system.
4. How did the mechanical advantage of your pulley system compare with those of your classmates?

Conclude and Apply

1. **Explain** how increasing the number of pulleys increases the mechanical advantage.
2. **Infer** How could you modify the pulley system to lift a weight twice as heavy with the same effort force used here?
3. **Compare** this real machine with an ideal machine.

Communicating Your Data

Show your design diagram to the class. Review the design and point out good and bad characteristics of your pulley system.
For more help, refer to the Science Skill Handbook.

LAB **M** ◆ 117

TIME SCIENCE AND Society
SCIENCE ISSUES THAT AFFECT YOU!

Bionic People

Artificial limbs can help people lead normal lives

People in need of transplants usually receive human organs. But many people's medical problems can only be solved by receiving artificial body parts. These synthetic devices, called prostheses, are used to replace anything from a heart valve to a knee joint. Bionics is the science of creating artificial body parts. A major focus of bionics is the replacement of lost limbs. Through accident, birth defect, or disease, people sometimes lack hands or feet, or even whole arms or legs.

For centuries, people have used prostheses to replace limbs. In the past, physically challenged people used devices like peg legs or artificial arms that ended in a pair of hooks. These prostheses didn't do much to replace lost functions of arms and legs.

The knowledge that muscles respond to electricity has helped create more effective prostheses. One such prostheses is the myoelectric arm. This battery-powered device connects muscle nerves in an amputated arm to a sensor.

The sensor detects when the arm tenses, then transmits the signal to an artificial hand, which opens or closes. New prosthetic hands even give a sense of touch, as well as cold and heat.

Myoelectric arms make life easier for people who have them.

Research Use your school's media center to find other aspects of robotics such as walking machines or robots that perform planetary exploration. What are they used for? How do they work? You could take it one step further and learn about cyborgs. Report to the class.

Science Online

For more information, visit bookm.msscience.com/time

Chapter 4 Study Guide

Reviewing Main Ideas

Section 1 Work and Power

1. Work is done when a force exerted on an object causes the object to move.
2. A force can do work only when it is exerted in the same direction as the object moves.
3. Work is equal to force times distance, and the unit of work is the joule.
4. Power is the rate at which work is done, and the unit of power is the watt.

Section 2 Using Machines

1. A machine can change the size or direction of an input force or the distance over which it is exerted.
2. The mechanical advantage of a machine is its output force divided by its input force.

Section 3 Simple Machines

1. A machine that does work with only one movement is a simple machine. A compound machine is a combination of simple machines.
2. Simple machines include the inclined plane, lever, wheel and axle, screw, wedge, and pulley.
3. Wedges and screws are inclined planes.
4. Pulleys can be used to multiply force and change direction.

Visualizing Main Ideas

Copy and complete the following concept map on simple machines.

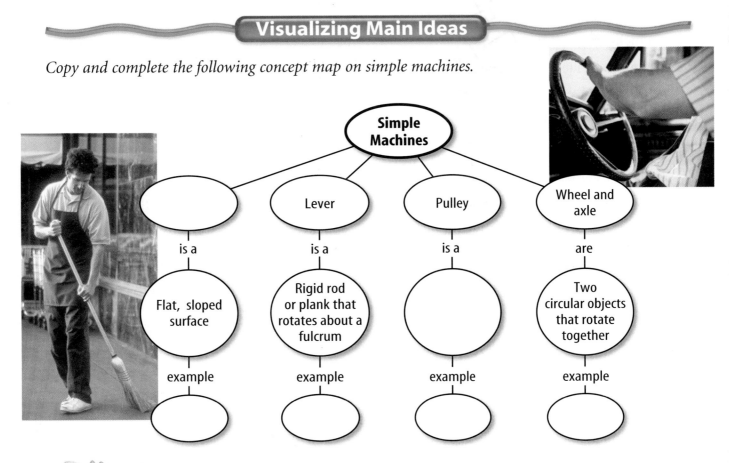

chapter 4 Review

Using Vocabulary

compound machine p. 109
efficiency p. 107
inclined plane p. 109
input force p. 104
lever p. 112
mechanical advantage p. 105
output force p. 104
power p. 101
pulley p. 114
screw p. 111
simple machine p. 109
wedge p. 110
wheel and axle p. 112
work p. 98

Each phrase below describes a vocabulary word. Write the vocabulary word that matches the phrase describing it.

1. percentage of work in to work out
2. force put into a machine
3. force exerted by a machine
4. two rigidly attached wheels
5. input force divided by output force
6. a machine with only one movement
7. an inclined plane that moves
8. a rigid rod that rotates about a fulcrum
9. a flat, sloped surface
10. amount of work divided by time

Checking Concepts

Choose the word or phrase that best answers the question.

11. Which of the following is a requirement for work to be done?
    A) Force is exerted.
    B) Object is carried.
    C) Force moves an object.
    D) Machine is used.

12. How much work is done when a force of 30 N moves an object a distance of 3 m?
    A) 3 J
    B) 10 J
    C) 30 J
    D) 90 J

13. How much power is used when 600 J of work are done in 10 s?
    A) 6 W
    B) 60 W
    C) 600 W
    D) 610 W

14. Which is a simple machine?
    A) baseball bat
    B) bicycle
    C) can opener
    D) car

15. Mechanical advantage can be calculated by which of the following expressions?
    A) input force/output force
    B) output force/input force
    C) input work/output work
    D) output work/input work

16. What is the ideal mechanical advantage of a machine that changes only the direction of the input force?
    A) less than 1
    B) zero
    C) 1
    D) greater than 1

Use the illustration below to answer question 17.

17. What is the output force if the input force on the wheel is 100 N?
    A) 5 N
    B) 200 N
    C) 500 N
    D) 2,000 N

18. Which of the following is a form of the inclined plane?
    A) pulley
    B) screw
    C) wheel and axle
    D) lever

19. For a given input force, a ramp increases which of the following?
    A) height
    B) output force
    C) output work
    D) efficiency

chapter 4 Review

Thinking Critically

Use the illustration below to answer question 20.

20. **Evaluate** Would a 9-N force applied 2 m from the fulcrum lift the weight? Explain.

21. **Explain** why the output work for any machine can't be greater than the input work.

22. **Explain** A doorknob is an example of a wheel and axle. Explain why turning the knob is easier than turning the axle.

23. **Infer** On the Moon, the force of gravity is less than on Earth. Infer how the mechanical advantage of an inclined plane would change if it were on the Moon, instead of on Earth.

24. **Make and Use Graphs** A pulley system has a mechanical advantage of 5. Make a graph with the input force on the x-axis and the output force on the y-axis. Choose five different values of the input force, and plot the resulting output force on your graph.

Use the diagram below to answer question 25.

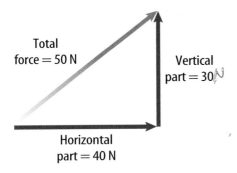

25. **Work** The diagram above shows a force exerted at an angle to pull a sled. How much work is done if the sled moves 10 m horizontally?

Performance Activities

26. **Identify** You have levers in your body. Your muscles and tendons provide the input force. Your joints act as fulcrums. The output force is used to move everything from your head to your hands. Describe and draw any human levers you can identify.

27. **Display** Make a display of everyday devices that are simple and compound machines. For devices that are simple machines, identify which simple machine it is. For compound machines, identify the simple machines that compose it.

Applying Math

28. **Mechanical Advantage** What is the mechanical advantage of a 6-m long ramp that extends from a ground-level sidewalk to a 2-m high porch?

29. **Input Force** How much input force is required to lift an 11,000-N beam using a pulley system with a mechanical advantage of 20?

30. **Efficiency** The input work done on a pulley system is 450 J. What is the efficiency of the pulley system if the output work is 375 J?

Use the table below to answer question 31.

Output Force Exerted by Machines		
Machine	Input Force (N)	Output Force (N)
A	500	750
B	300	200
C	225	225
D	800	1,100
E	75	110

31. **Mechanical Advantage** According to the table above, which of the machines listed has the largest mechanical advantage?

chapter 4 Standardized Test Practice

Part 1 Multiple Choice

Record your answers on the answer sheet provided by your teacher or on a sheet of paper.

1. The work done by a boy pulling a snow sled up a hill is 425 J. What is the power expended by the boy if he pulls on the sled for 10.5 s?
   A. 24.7 W C. 247 W
   B. 40.5 W D. 4460 W

Use the illustration below to answer questions 2 and 3.

2. What is the mechanical advantage of the lever shown above?
   A. $\frac{1}{6}$ C. 2
   B. $\frac{1}{2}$ D. 6

3. What would the mechanical advantage of the lever be if the triangular block were moved to a position 35 cm from the edge of the output force side of the plank?
   A. $\frac{1}{4}$ C. 3
   B. $\frac{1}{3}$ D. 4

4. Which of the following causes the efficiency of a machine to be less than 100%?
   A. work
   B. power
   C. mechanical advantage
   D. friction

Test-Taking Tip

Simplify Diagrams Write directly on complex charts, such as a Punnett square.

Use the illustration below to answer questions 5 and 6.

5. The pulley system in the illustration above uses several pulleys to increase the mechanical advantage. What is the mechanical advantage of this system?
   A. 1 C. 3
   B. 2 D. 4

6. Suppose the lower pulley was removed so that the object was supported only by the upper pulley. What would the mechanical advantage be?
   A. 0 C. 2
   B. 1 D. 3

7. You push a shopping cart with a force of 12 N for a distance of 1.5 m. You stop pushing the cart, but it continues to roll for 1.1 m. How much work did you do?
   A. 8.0 J C. 18 J
   B. 13 J   D. 31 J

8. What is the mechanical advantage of a wheel with a radius of 8.0 cm connected to an axle with a radius of 2.5 cm?
   A. 0.31 C. 3.2
   B. 2.5   D. 20

9. You push a 5-kg box across the floor with a force of 25 N. How far do you have to push the box to do 63 J of work?
   A. 0.40 m C. 2.5 m
   B. 1.6 m   D. 13 m

122 ◆ M STANDARDIZED TEST PRACTICE

Standardized Test Practice

Part 2 Short Response/Grid In

Record your answers on the answer sheet provided by your teacher or on a sheet of paper.

10. What is the name of the point about which a lever rotates?

11. Describe how you can determine the mechanical advantage of a pulley or a pulley system.

Use the figure below to answer questions 12 and 13.

12. What type of simple machine is the tip of the dart in the photo above?

13. Would the mechanical advantage of the dart tip change if the tip were longer and thinner? Explain.

14. How much energy is used by a 75-W lightbulb in 15 s?

15. The input and output forces are applied at the ends of the lever. If the lever is 3 m long and the output force is applied 1 m from the fulcrum, what is the mechanical advantage?

16. Your body contains simple machines. Name one part that is a wedge and one part that is a lever.

17. Explain why applying a lubricant, such as oil, to the surfaces of a machine causes the efficiency of the machine to increase.

18. Apply the law of conservation of energy to explain why the output work done by a real machine is always less than the input work done on the machine.

Part 3 Open Ended

Record your answers on a sheet of paper.

19. The output work of a machine can never be greater than the input work. However, the advantage of using a machine is that it makes work easier. Describe and give an example of the three ways a machine can make work easier.

20. A wheel and axle may have a mechanical advantage that is either greater than 1 or less than 1. Describe both types and give some examples of each.

21. Draw a sketch showing the cause of friction as two surfaces slide past each other. Explain your sketch, and describe how lubrication can reduce the friction between the two surfaces.

22. Draw the two types of simple pulleys and an example of a combination pulley. Draw arrows to show the direction of force on your sketches.

Use the figure below to answer question 23.

23. Identify two simple machines in the photo above and describe how they make riding a bicycle easier.

24. Explain why the mechanical advantage of an inclined plane can never be less than 1.

bookm.msscience.com/standardized_test

chapter 5

Energy and Energy Resources

The BIG Idea

Changes occur when energy is transferred from one place to another.

SECTION 1
What is energy?
Main Idea Energy can have different forms.

SECTION 2
Energy Transformations
Main Idea Energy can change form, but can never be created or destroyed.

SECTION 3
Sources of Energy
Main Idea Energy contained in different energy sources is transformed into useful forms of energy.

Blowing Off Steam

The electrical energy you used today might have been produced by a coal-burning power plant like this one. Energy contained in coal is transformed into heat, and then into electrical energy. As boiling water heated by the burning coal is cooled, steam rises from these cone-shaped cooling towers.

Science Journal Choose three devices that use electricity, and identify the function of each device.

Start-Up Activities

Marbles and Energy

What's the difference between a moving marble and one at rest? A moving marble can hit something and cause a change to occur. How can a marble acquire energy—the ability to cause change?

1. Make a track on a table by slightly separating two metersticks placed side by side.
2. Using a book, raise one end of the track slightly and measure the height.
3. Roll a marble down the track. Measure the distance from its starting point to where it hits the floor. Repeat. Calculate the average of the two measurements.
4. Repeat steps 2 and 3 for three different heights. Predict what will happen if you use a heavier marble. Test your prediction and record your observations.
5. **Think Critically** In your Science Journal, describe how the distance traveled by the marble is related to the height of the ramp. How is the motion of the marble related to the ramp height?

Energy Make the following Foldable to help identify what you already know, what you want to know, and what you learned about energy.

STEP 1 Fold a vertical sheet of paper from side to side. Make the front edge about 1 cm shorter than the back edge.

STEP 2 Turn lengthwise and fold into thirds

STEP 3 Unfold, cut, and label each tab for only the top layer along both folds to make three tabs.

Identify Questions Before you read the chapter, write what you know and what you want to know about the types, sources, and transformation of energy under the appropriate tabs. As you read the chapter, correct what you have written and add more questions under the *Learned* tab.

Preview this chapter's content and activities at
bookm.msscience.com

Get Ready to Read

Take Notes

① Learn It! The best way for you to remember information is to write it down, or take notes. Good note-taking is useful for studying and research. When you are taking notes, it is helpful to
- phrase the information in your own words;
- restate ideas in short, memorable phrases;
- stay focused on main ideas and only the most important supporting details.

② Practice It! Make note-taking easier by using a chart to help you organize information clearly. Write the main ideas in the left column. Then write at least three supporting details in the right column. Read the text from Section 2 of this chapter under the heading *Energy Changes Form*, pages 133–135. Then take notes using a chart, such as the one below.

Main Idea	Supporting Details
	1. 2. 3. 4. 5.
	1. 2. 3. 4. 5.

③ Apply It! As you read this chapter, make a chart of the main ideas. Next to each main idea, list at least two supporting details.

Target Your Reading

Reading Tip
Read one or two paragraphs first and take notes after you read. You are likely to take down too much information if you take notes as you read.

Use this to focus on the main ideas as you read the chapter.

① Before you read the chapter, respond to the statements below on your worksheet or on a numbered sheet of paper.
- Write an **A** if you **agree** with the statement.
- Write a **D** if you **disagree** with the statement.

② After you read the chapter, look back to this page to see if you've changed your mind about any of the statements.
- If any of your answers changed, explain why.
- Change any false statements into true statements.
- Use your revised statements as a study guide.

Science Online
Print out a worksheet of this page at bookm.msscience.com

Before You Read A or D		Statement	After You Read A or D
	1	Only objects that are moving have energy.	
	2	Energy of motion depends only on the object's mass and speed.	
	3	The amount of thermal energy an object has depends on its temperature.	
	4	Energy changes from one form to another when a ball is thrown upward.	
	5	Living organisms change chemical energy to thermal energy.	
	6	The total amount of energy increases when a candle burns.	
	7	Coal is being made faster than it is being used.	
	8	Solar energy is the least expensive energy source.	
	9	A heat pump can heat and cool a building.	

section 1

What is energy?

as you read

What You'll Learn
- **Explain** what energy is.
- **Distinguish** between kinetic energy and potential energy.
- **Identify** the various forms of energy.

Why It's Important
Energy is involved whenever a change occurs.

Review Vocabulary
mass: a measure of the amount of matter in an object

New Vocabulary
- energy
- kinetic energy
- potential energy
- thermal energy
- chemical energy
- radiant energy
- electrical energy
- nuclear energy

The Nature of Energy

What comes to mind when you hear the word *energy*? Do you picture running, leaping, and spinning like a dancer or a gymnast? How would you define energy? When an object has energy, it can make things happen. In other words, **energy** is the ability to cause change. What do the items shown in **Figure 1** have in common?

Look around and notice the changes that are occurring—someone walking by or a ray of sunshine that is streaming through the window and warming your desk. Maybe you can see the wind moving the leaves on a tree. What changes are occurring?

Transferring Energy You might not realize it, but you have a large amount of energy. In fact, everything around you has energy, but you notice it only when a change takes place. Anytime a change occurs, energy is transferred from one object to another. You hear a footstep because energy is transferred from a foot hitting the ground to your ears. Leaves are put into motion when energy in the moving wind is transferred to them. The spot on the desktop becomes warmer when energy is transferred to it from the sunlight. In fact, all objects, including leaves and desktops, have energy.

Figure 1 Energy is the ability to cause change.
Explain how these objects cause change.

Energy of Motion

Things that move can cause change. A bowling ball rolls down the alley and knocks down some pins, as in **Figure 2A.** Is energy involved? A change occurs when the pins fall over. The bowling ball causes this change, so the bowling ball has energy. The energy in the motion of the bowling ball causes the pins to fall. As the ball moves, it has a form of energy called kinetic energy. **Kinetic energy** is the energy an object has due to its motion. If an object isn't moving, it doesn't have kinetic energy.

Figure 2 The kinetic energy of an object depends on the mass and speed of the object.

Kinetic Energy and Speed If you roll the bowling ball so it moves faster, what happens when it hits the pins? It might knock down more pins, or it might cause the pins to go flying farther. A faster ball causes more change to occur than a ball that is moving slowly. Look at **Figure 2B.** The professional bowler rolls a fast-moving bowling ball. When her ball hits the pins, pins go flying faster and farther than for a slower-moving ball. All that action signals that her ball has more energy. The faster the ball goes, the more kinetic energy it has. This is true for all moving objects. Kinetic energy increases as an object moves faster.

A This ball has kinetic energy because it is rolling down the alley.

Reading Check *How does kinetic energy depend on speed?*

B This ball has more kinetic energy because it has more speed.

Kinetic Energy and Mass Suppose, as shown in **Figure 2C,** you roll a volleyball down the alley instead of a bowling ball. If the volleyball travels at the same speed as a bowling ball, do you think it will send pins flying as far? The answer is no. The volleyball might not knock down any pins. Does the volleyball have less energy than the bowling ball even though they are traveling at the same speed?

An important difference between the volleyball and the bowling ball is that the volleyball has less mass. Even though the volleyball is moving at the same speed as the bowling ball, the volleyball has less kinetic energy because it has less mass. Kinetic energy also depends on the mass of a moving object. Kinetic energy increases as the mass of the object increases.

C This ball has less kinetic energy because it has less mass.

Figure 3 The potential energy of an object depends on its mass and height above the ground.
Determine which vase has more potential energy, the red one or the blue one.

Energy of Position

An object can have energy even though it is not moving. For example, a glass of water sitting on the kitchen table doesn't have any kinetic energy because it isn't moving. If you accidentally nudge the glass and it falls on the floor, changes occur. Gravity pulls the glass downward, and the glass has energy of motion as it falls. Where did this energy come from?

When the glass was sitting on the table, it had potential (puh TEN chul) energy. **Potential energy** is the energy stored in an object because of its position. In this case, the position is the height of the glass above the floor. The potential energy of the glass changes to kinetic energy as the glass falls. The potential energy of the glass is greater if it is higher above the floor. Potential energy also depends on mass. The more mass an object has, the more potential energy it has. Which object in **Figure 3** has the most potential energy?

Forms of Energy

Food, sunlight, and wind have energy, yet they seem different because they contain different forms of energy. Food and sunlight contain forms of energy different from the kinetic energy in the motion of the wind. The warmth you feel from sunlight is another type of energy that is different from the energy of motion or position.

Thermal Energy The feeling of warmth from sunlight signals that your body is acquiring more thermal energy. All objects have **thermal energy** that increases as its temperature increases. A cup of hot chocolate has more thermal energy than a cup of cold water, as shown in **Figure 4.** Similarly, the cup of water has more thermal energy than a block of ice of the same mass. Your body continually produces thermal energy. Many chemical reactions that take place inside your cells produce thermal energy. Where does this energy come from? Thermal energy released by chemical reactions comes from another form of energy called chemical energy.

Figure 4 The hotter an object is, the more thermal energy it has. A cup of hot chocolate has more thermal energy than a cup of cold water, which has more thermal energy than a block of ice with the same mass.

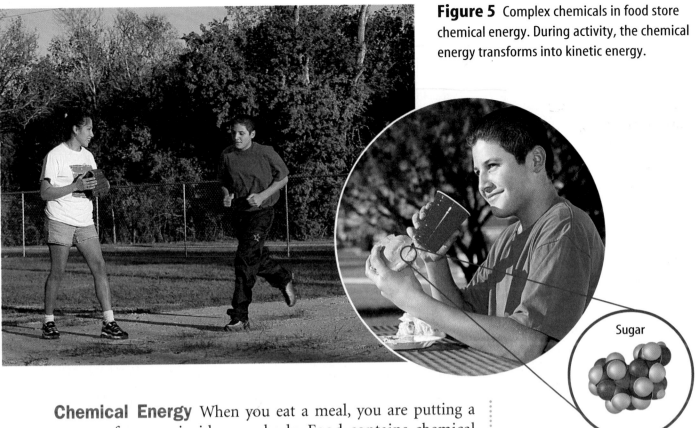

Figure 5 Complex chemicals in food store chemical energy. During activity, the chemical energy transforms into kinetic energy.

Chemical Energy When you eat a meal, you are putting a source of energy inside your body. Food contains chemical energy that your body uses to provide energy for your brain, to power your movements, and to fuel your growth. As in **Figure 5,** food contains chemicals, such as sugar, which can be broken down in your body. These chemicals are made of atoms that are bonded together, and energy is stored in the bonds between atoms. **Chemical energy** is the energy stored in chemical bonds. When chemicals are broken apart and new chemicals are formed, some of this energy is released. The flame of a candle is the result of chemical energy stored in the wax. When the wax burns, chemical energy is transformed into thermal energy and light energy.

Reading Check *When is chemical energy released?*

Light Energy Light from the candle flame travels through the air at an incredibly fast speed of 300,000 km/s. This is fast enough to circle Earth almost eight times in 1 s. When light strikes something, it can be absorbed, transmitted, or reflected. When the light is absorbed by an object, the object can become warmer. The object absorbs energy from the light and this energy is transformed into thermal energy. Then energy carried by light is called **radiant energy**. **Figure 6** shows a coil of wire that produces radiant energy when it is heated. To heat the metal, another type of energy can be used—electrical energy.

Figure 6 Electrical energy is transformed into thermal energy in the metal heating coil. As the metal becomes hotter, it emits more radiant energy.

SECTION 1 What is energy? **M** ◆ **129**

Electrical Energy Electrical lighting is one of the many ways electrical energy is used. Look around at all the devices that use electricity. Electric current flows in these devices when they are connected to batteries or plugged into an electric outlet. **Electrical energy** is the energy that is carried by an electric current. An electric device uses the electrical energy provided by the current flowing in the device. Large electric power plants generate the enormous amounts of electrical energy used each day. About 20 percent of the electrical energy used in the United States is generated by nuclear power plants.

Nuclear Energy Nuclear power plants use the energy stored in the nucleus of an atom to generate electricity. Every atomic nucleus contains energy—**nuclear energy**—that can be transformed into other forms of energy. However, releasing the nuclear energy is a difficult process. It involves the construction of complex power plants, shown in **Figure 7**. In contrast, all that is needed to release chemical energy from wood is a lighted match.

Figure 7 Complex power plants are required to obtain useful energy from the nucleus of an atom.

section 1 review

Summary

The Nature of Energy
- Energy is the ability to cause change.
- Kinetic energy is the energy an object has due to its motion. Kinetic energy depends on an object's speed and mass.
- Potential energy is the energy an object has due to its position. Potential energy depends on an object's height and mass.

Forces of Energy
- Thermal energy increases as temperature increases.
- Chemical energy is the energy stored in chemical bonds in molecules.
- Light energy, also called radiant energy, is the energy contained in light.
- Electrical energy is the energy carried by electric current.
- Nuclear energy is the energy contained in the nucleus of an atom.

Self Check

1. **Explain** why a high-speed collision between two cars would cause more damage than a low-speed collision between the same two cars.
2. **Describe** the energy transformations that occur when a piece of wood is burned.
3. **Identify** the form of energy that is converted into thermal energy by your body.
4. **Explain** how, if two vases are side by side on a shelf, one could have more potential energy.
5. **Think Critically** A golf ball and a bowling ball are moving and both have the same kinetic energy. Which one is moving faster? If they move at the same speed, which one has more kinetic energy?

Applying Skills

6. **Communicate** In your Science Journal, record different ways the word *energy* is used. Which ways of using the word *energy* are closest to the definition of energy given in this section?

bookm.msscience.com/self_check_quiz

section 2

Energy Transformations

Changing Forms of Energy

Chemical, thermal, radiant, and electrical are some of the forms that energy can have. In the world around you, energy is transforming continually between one form and another. You observe some of these transformations by noticing a change in your environment. Forest fires are a dramatic example of an environmental change that can occur naturally as a result of lightning strikes. A number of changes occur that involve energy as the mountain biker in **Figure 8** pedals up a hill. What energy transformations cause these changes to occur?

Tracking Energy Transformations As the mountain biker pedals, his leg muscles transform chemical energy into kinetic energy. The kinetic energy of his leg muscles transforms into kinetic energy of the bicycle as he pedals. Some of this energy transforms into potential energy as he moves up the hill. Also, some energy is transformed into thermal energy. His body is warmer because chemical energy is being released. Because of friction, the mechanical parts of the bicycle are warmer, too. Thermal energy is almost always produced by an energy transformation. The energy transformations that occur when people exercise, when cars run, when living things grow and even when stars explode, all produce thermal energy.

as you read

What You'll Learn
- **Apply** the law of conservation of energy to energy transformations.
- **Identify** how energy changes form.
- **Describe** how electric power plants produce energy.

Why It's Important
Changing energy from one form to another is what makes cars run, furnaces heat, telephones work, and plants grow.

Review Vocabulary
transformation: a change in composition or structure

New Vocabulary
- law of conservation of energy
- generator
- turbine

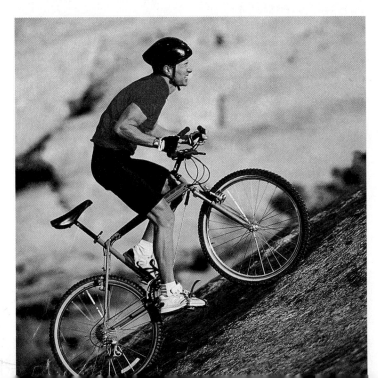

Figure 8 The ability to transform energy allows the biker to climb the hill.
Identify *all the forms of energy that are represented in the photograph.*

M ◆ 131

Topic: Energy Transformations

Visit bookm.msscience.com for Web links to information about energy transformations that occur during different activities and processes.

Activity Choose an activity or process and make a graph showing how the kinetic and potential energy change during it.

The Law of Conservation of Energy

It can be a challenge to track energy as it moves from object to object. However, one extremely important principle can serve as a guide as you trace the flow of energy. According to the **law of conservation of energy,** energy is never created or destroyed. The only thing that changes is the form in which energy appears. When the biker is resting at the summit, all his original energy is still around. Some of the energy is in the form of potential energy, which he will use as he coasts down the hill. Some of this energy was changed to thermal energy by friction in the bike. Chemical energy was also changed to thermal energy in the biker's muscles, making him feel hot. As he rests, this thermal energy moves from his body to the air around him. No energy is missing—it can all be accounted for.

Reading Check *Can energy ever be lost? Why or why not?*

Changing Kinetic and Potential Energy

The law of conservation of energy can be used to identify the energy changes in a system. For example, tossing a ball into the air and catching it is a simple system. As shown in **Figure 9,** as the ball leaves your hand, most of its energy is kinetic. As the ball rises, it slows and its kinetic energy decreases. But, the total energy of the ball hasn't changed. The decrease in kinetic energy equals the increase in potential energy as the ball flies higher in the air. The total amount of energy remains constant. Energy moves from place to place and changes form, but it never is created or destroyed.

Figure 9 During the flight of the baseball, energy is transforming between kinetic and potential energy.
Determine *where the ball has the most kinetic energy. Where does the ball have the most total energy?*

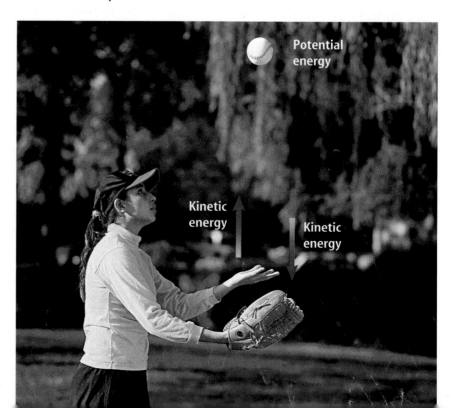

Figure 10 Hybrid cars that use an electric motor and a gasoline engine for power are now available. Hybrid cars make energy transformations more efficient.

Energy Changes Form

Energy transformations occur constantly all around you. Many machines are devices that transform energy from one form to another. For example, an automobile engine transforms the chemical energy in gasoline into energy of motion. However, not all of the chemical energy is converted into kinetic energy. Instead, some of the chemical energy is converted into thermal energy, and the engine becomes hot. An engine that converts chemical energy into more kinetic energy is a more efficient engine. New types of cars, like the one shown in **Figure 10,** use an electric motor along with a gasoline engine. These engines are more efficient so the car can travel farther on a gallon of gas.

Transforming Chemical Energy Inside your body, chemical energy also is transformed into kinetic energy. Look at **Figure 11.** The transformation of chemical to kinetic energy occurs in muscle cells. There, chemical reactions take place that cause certain molecules to change shape. Your muscle contracts when many of these changes occur, and a part of your body moves.

The matter contained in living organisms, also called biomass, contains chemical energy. When organisms die, chemical compounds in their biomass break down. Bacteria, fungi, and other organisms help convert these chemical compounds to simpler chemicals that can be used by other living things.

Thermal energy also is released as these changes occur. For example, a compost pile can contain plant matter, such as grass clippings and leaves. As the compost pile decomposes, chemical energy is converted into thermal energy. This can cause the temperature of a compost pile to reach 60°C.

Analyzing Energy Transformations

Procedure
1. Place soft **clay** on the floor and smooth out its surface.
2. Hold a **marble** 1.5 m above the clay and drop it. Measure the depth of the crater made by the marble.
3. Repeat this procedure using a **golf ball** and a **plastic golf ball.**

Analysis
1. Compare the depths of the craters to determine which ball had the most kinetic energy as it hit the clay.
2. Explain how potential energy was transformed into kinetic energy during your activity.

SECTION 2 Energy Transformations **M** ◆ **133**

NATIONAL GEOGRAPHIC VISUALIZING ENERGY TRANSFORMATIONS

Figure 11

Paddling a raft, throwing a baseball, playing the violin — your skeletal muscles make these and countless other body movements possible. Muscles work by pulling, or contracting. At the cellular level, muscle contractions are powered by reactions that transform chemical energy into kinetic energy.

▶ Energy transformations taking place in your muscles provide the power to move.

▲ Many skeletal muscles are arranged in pairs that work in opposition to each other. When you bend your arm, the biceps muscle contracts, while the triceps relaxes. When you extend your arm the triceps contracts, and the biceps relaxes.

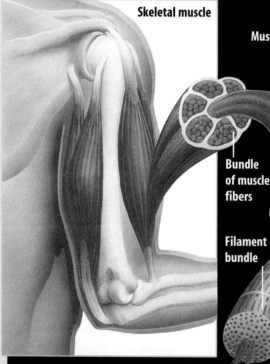

▲ Skeletal muscles are made up of bundles of muscle cells, or fibers. Each fiber is composed of many bundles of muscle filaments.

▲ A signal from a nerve fiber starts a chemical reaction in the muscle filament. This causes molecules in the muscle filament to gain energy and move. Many filaments moving together cause the muscle to contract.

Figure 12 The simple act of listening to a radio involves many energy transformations. A few are diagrammed here.

Transforming Electrical Energy Every day you use electrical energy. When you flip a light switch, or turn on a radio or television, or use a hair drier, you are transforming electrical energy to other forms of energy. Every time you plug something into a wall outlet, or use a battery, you are using electrical energy. **Figure 12** shows how electrical energy is transformed into other forms of energy when you listen to a radio. A loudspeaker in the radio converts electrical energy into sound waves that travel to your ear—energy in motion. The energy that is carried by the sound waves causes parts of the ear to move also. This energy of motion is transformed again into chemical and electrical energy in nerve cells, which send the energy to your brain. After your brain interprets this energy as a voice or music, where does the energy go? The energy finally is transformed into thermal energy.

Transforming Thermal Energy Different forms of energy can be transformed into thermal energy. For example, chemical energy changes into thermal energy when something burns. Electrical energy changes into thermal energy when a wire that is carrying an electric current gets hot. Thermal energy can be used to heat buildings and keep you warm. Thermal energy also can be used to heat water. If water is heated to its boiling point, it changes to steam. This steam can be used to produce kinetic energy by steam engines, like the steam locomotives that used to pull trains. Thermal energy also can be transformed into radiant energy. For example, when a bar of metal is heated to a high temperature, it glows and gives off light.

Controlling Body Temperature Most organisms have some adaptation for controlling the amount of thermal energy in their bodies. Some living in cooler climates have thick fur coats that help prevent thermal energy from escaping, and some living in desert regions have skin that helps keep thermal energy out. Research some of the adaptations different organisms have for controlling the thermal energy in their bodies.

How Thermal Energy Moves Thermal energy can move from one place to another. Look at **Figure 13.** The hot chocolate has thermal energy that moves from the cup to the cooler air around it, and to the cooler spoon. Thermal energy only moves from something at a higher temperature to something at a lower temperature.

Generating Electrical Energy

The enormous amount of electrical energy that is used every day is too large to be stored in batteries. The electrical energy that is available for use at any wall socket must be generated continually by power plants. Every power plant works on the same principle—energy is used to turn a large generator. A **generator** is a device that transforms kinetic energy into electrical energy. In fossil fuel power plants, coal, oil, or natural gas is burned to boil water. As the hot water boils, the steam rushes through a **turbine,** which contains a set of narrowly spaced fan blades. The steam pushes on the blades and turns the turbine, which in turn rotates a shaft in the generator to produce the electrical energy, as shown in **Figure 14.**

Figure 13 Thermal energy moves from the hot chocolate to the cooler surroundings. **Explain** what happens to the hot chocolate as it loses thermal energy.

Figure 14 A coal-burning power plant transforms the chemical energy in coal into electrical energy. **List** some of the other energy sources that power plants use.

Reading Check *What does a generator do?*

Power Plants Almost 90 percent of the electrical energy generated in the United States is produced by nuclear and fossil fuel power plants, as shown in **Figure 15.** Other types of power plants include hydroelectric (hi droh ih LEK trihk) and wind. Hydroelectric power plants transform the kinetic energy of moving water into electrical energy. Wind power plants transform the kinetic energy of moving air into electrical energy. In these power plants, a generator converts the kinetic energy of moving water or wind to electrical energy.

To analyze the energy transformations in a power plant, you can diagram the energy changes using arrows. A coal-burning power plant generates electrical energy through the following series of energy transformations.

chemical energy of coal → thermal energy of water → kinetic energy of steam → kinetic energy of turbine → electrical energy out of generator

Nuclear power plants use a similar series of transformations. Hydroelectric plants, however, skip the steps that change water into steam because the water strikes the turbine directly.

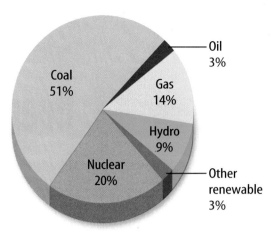

Figure 15 The graph shows sources of electrical energy in the United States.
Name *the energy source that you think is being used to provide the electricity for the lights overhead.*

section 2 review

Summary

Changing Forms of Energy
- Heat usually is one of the forms of energy produced in energy transformations.
- The law of conservation of energy states that energy cannot be created or destroyed; it can only change form.
- The total energy doesn't change when an energy transformation occurs.
- As an object rises and falls, kinetic and potential energy are transformed into each other, but the total energy doesn't change.

Generating Electrical Energy
- A generator converts kinetic energy into electrical energy.
- Burning fossil fuels produces thermal energy that is used to boil water and produce steam.
- In a power plant, steam is used to spin a turbine which then spins an electric generator.

Self Check

1. **Describe** the conversions between potential and kinetic energy that occur when you shoot a basketball at a basket.
2. **Explain** whether your body gains or loses thermal energy if your body temperature is 37°C and the temperature around you is 25°C.
3. **Describe** a process that converts chemical energy to thermal energy.
4. **Think Critically** A lightbulb converts 10 percent of the electrical energy it uses into radiant energy. Make a hypothesis about the other form of energy produced.

Applying Math

5. **Use a Ratio** How many times greater is the amount of electrical energy produced in the United States by coal-burning power plants than the amount produced by nuclear power plants?

Hearing with Your Jaw

You probably have listened to music using speakers or headphones. Have you ever considered how energy is transferred to get the energy from the radio or CD player to your brain? What type of energy is needed to power the radio or CD player? Where does this energy come from? How does that energy become sound? How does the sound get to you? In this activity, the sound from a radio or CD player is going to travel through a motor before entering your body through your jaw instead of your ears.

● Real-World Question

How can energy be transferred from a radio or CD player to your brain?

Goals
- **Identify** energy transfers and transformations.
- **Explain** your observations using the law of conservation of energy.

Materials
radio or CD player
small electrical motor
headphone jack

● Procedure

1. Go to one of the places in the room with a motor/radio assembly.
2. Turn on the radio or CD player so that you hear the music.
3. Push the headphone jack into the headphone plug on the radio or CD player.
4. Press the axle of the motor against the side of your jaw.

● Conclude and Apply

1. **Describe** what you heard in your Science Journal.
2. **Identify** the form of energy produced by the radio or CD player.
3. **Draw** a diagram to show all of the energy transformations taking place.
4. **Evaluate** Did anything get hotter as a result of this activity? Explain.
5. **Explain** your observations using the law of conservation of energy.

Communicating Your Data

Compare your conclusions with those of other students in your class. **For more help, refer to the** Science Skill Handbook.

138 ◆ **M CHAPTER 5** Energy and Energy Resources

section 3
Sources of Energy

Using Energy

Every day, energy is used to provide light and to heat and cool homes, schools, and workplaces. According to the law of conservation of energy, energy can't be created or destroyed. Energy only can change form. If a car or refrigerator can't create the energy they use, then where does this energy come from?

Energy Resources

Energy cannot be made, but must come from the natural world. As you can see in **Figure 16,** the surface of Earth receives energy from two sources—the Sun and radioactive atoms in Earth's interior. The amount of energy Earth receives from the Sun is far greater than the amount generated in Earth's interior. Nearly all the energy you used today can be traced to the Sun, even the gasoline used to power the car or school bus you came to school in.

as you read

What You'll Learn
- **Explain** what renewable, non-renewable, and alternative resources are.
- **Describe** the advantages and disadvantages of using various energy sources.

Why It's Important
Energy is vital for survival and making life comfortable. Developing new energy sources will improve modern standards of living.

Review Vocabulary
resource: a natural feature or phenomenon that enhances the quality of life

New Vocabulary
- nonrenewable resource
- renewable resource
- alternative resource
- inexhaustible resource
- photovoltaic

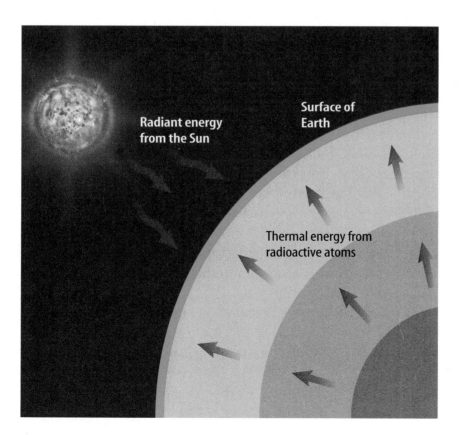

Figure 16 All the energy you use can be traced to one of two sources—the Sun or radioactive atoms in Earth's interior.

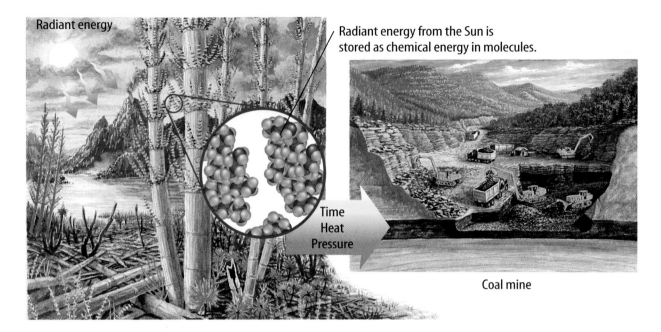

Figure 17 Coal is formed after the molecules in ancient plants are heated under pressure for millions of years. The energy stored by the molecules in coal originally came from the Sun.

Energy Source Origins The kinds of fossil fuels found in the ground depend on the kinds of organisms (animal or plant) that died and were buried in that spot. Research coal, oil, and natural gas to find out what types of organisms were primarily responsible for producing each.

Fossil Fuels

Fossil fuels are coal, oil, and natural gas. Oil and natural gas were made from the remains of microscopic organisms that lived in Earth's oceans millions of years ago. Heat and pressure gradually turned these ancient organisms into oil and natural gas. Coal was formed by a similar process from the remains of ancient plants that once lived on land, as shown in **Figure 17.**

Through the process of photosynthesis, ancient plants converted the radiant energy in sunlight to chemical energy stored in various types of molecules. Heat and pressure changed these molecules into other types of molecules as fossil fuels formed. Chemical energy stored in these molecules is released when fossil fuels are burned.

Using Fossil Fuels The energy used when you ride in a car, turn on a light, or use an electric appliance usually comes from burning fossil fuels. However, it takes millions of years to replace each drop of gasoline and each lump of coal that is burned. This means that the supply of oil on Earth will continue to decrease as oil is used. An energy source that is used up much faster than it can be replaced is a **nonrenewable resource.** Fossil fuels are nonrenewable resources.

Burning fossil fuels to produce energy also generates chemical compounds that cause pollution. Each year billions of kilograms of air pollutants are produced by burning fossil fuels. These pollutants can cause respiratory illnesses and acid rain. Also, the carbon dioxide gas formed when fossil fuels are burned might cause Earth's climate to warm.

Nuclear Energy

Can you imagine running an automobile on 1 kg of fuel that releases almost 3 million times more energy than 1 L of gas? What could supply so much energy from so little mass? The answer is the nuclei of uranium atoms. Some of these nuclei are unstable and break apart, releasing enormous amounts of energy in the process. This energy can be used to generate electricity by heating water to produce steam that spins an electric generator, as shown in **Figure 18.** Because no fossil fuels are burned, generating electricity using nuclear energy helps make the supply of fossil fuels last longer. Also, unlike fossil fuel power plants, nuclear power plants produce almost no air pollution. In one year, a typical nuclear power plant generates enough energy to supply 600,000 homes with power and produces only 1 m^3 of waste.

Nuclear Wastes Like all energy sources, nuclear energy has its advantages and disadvantages. One disadvantage is the amount of uranium in Earth's crust is nonrenewable. Another is that the waste produced by nuclear power plants is radioactive and can be dangerous to living things. Some of the materials in the nuclear waste will remain radioactive for many thousands of years. As a result the waste must be stored so no radioactivity is released into the environment for a long time. One method is to seal the waste in a ceramic material, place the ceramic in protective containers, and then bury the containers far underground. However, the burial site would have to be chosen carefully so underground water supplies aren't contaminated. Also, the site would have to be safe from earthquakes and other natural disasters that might cause radioactive material to be released.

Figure 18 To obtain electrical energy from nuclear energy, a series of energy transformations must occur.

Hydroelectricity

Currently, transforming the potential energy of water that is trapped behind dams supplies the world with almost 20 percent of its electrical energy. Hydroelectricity is the largest renewable source of energy. A **renewable resource** is an energy source that is replenished continually. As long as enough rain and snow fall to keep rivers flowing, hydroelectric power plants can generate electrical energy, as shown in **Figure 19.**

Although production of hydroelectricity is largely pollution free, it has one major problem. It disrupts the life cycle of aquatic animals, especially fish. This is particularly true in the Northwest where salmon spawn and run. Because salmon return to the spot where they were hatched to lay their eggs, the development of dams has hindered a large fraction of salmon from reproducing. This has greatly reduced the salmon population. Efforts to correct the problem have resulted in plans to remove a number of dams. In an attempt to help fish bypass some dams, fish ladders are being installed. Like most energy sources, hydroelectricity has advantages and disadvantages.

Topic: Hydroelectricity
Visit bookm.msscience.com for Web links to information about the use of hydroelectricity in various parts of the world.

Activity On a map of the world, show where the use of hydroelectricity is the greatest.

Applying Science

Is energy consumption outpacing production?

You use energy every day—to get to school, to watch TV, and to heat or cool your home. The amount of energy consumed by an average person has increased over time. Consequently, more energy must be produced.

Identifying the Problem

The graph above shows the energy produced and consumed in the United States from 1949 to 1999. How does energy that is consumed by Americans compare with energy that is produced in the United States?

Solving the Problem

1. Determine the approximate amount of energy produced in 1949 and in 1999 and how much it has increased in 50 years. Has it doubled or tripled?
2. Do the same for consumption. Has it doubled or tripled?
3. Using your answers for steps 1 and 2 and the graph, where does the additional energy that is needed come from? Give some examples.

Figure 19 The potential energy of water behind a dam supplies the energy to turn the turbine. **Explain** *why hydroelectric power is a renewable energy source.*

Alternative Sources of Energy

Electrical energy can be generated in several ways. However, each has disadvantages that can affect the environment and the quality of life for humans. Research is being done to develop new sources of energy that are safer and cause less harm to the environment. These sources often are called **alternative resources.** These alternative resources include solar energy, wind, and geothermal energy.

Solar Energy

The Sun is the origin of almost all the energy that is used on Earth. Because the Sun will go on producing an enormous amount of energy for billions of years, the Sun is an inexhaustible source of energy. An **inexhaustible resource** is an energy source that can't be used up by humans.

Each day, on average, the amount of solar energy that strikes the United States is more than the total amount of energy used by the entire country in a year. However, less than 0.1 percent of the energy used in the United States comes directly from the Sun. One reason is that solar energy is more expensive to use than fossil fuels. However, as the supply of fossil fuels decreases, the cost of finding and mining these fuels might increase. Then, it may be cheaper to use solar energy or other energy sources to generate electricity and heat buildings than to use fossil fuels.

 What is an inexhaustible energy source?

Building a Solar Collector

Procedure
1. Line a **large pot** with **black plastic** and fill with **water.**
2. Stretch **clear-plastic wrap** over the pot and tape it taut.
3. Make a slit in the top and slide a **thermometer** or a **computer probe** into the water.
4. Place your solar collector in direct sunlight and monitor the temperature change every 3 min for 15 min.
5. Repeat your experiment without using any black plastic.

Analysis
1. Graph the temperature changes in both setups.
2. Explain how your solar collector works.

SECTION 3 Sources of Energy **M ◆ 143**

Collecting the Sun's Energy Two types of collectors capture the Sun's rays. If you look around your neighborhood, you might see large, rectangular panels attached to the roofs of buildings or houses. If, as in **Figure 20,** pipes come out of the panel, it is a thermal collector. Using a black surface, a thermal collector heats water by directly absorbing the Sun's radiant energy. Water circulating in this system can be heated to about 70°C. The hot water can be pumped through the house to provide heat. Also, the hot water can be used for washing and bathing. If the panel has no pipes, it is a photovoltaic (foh toh vol TAY ihk) collector, like the one pictured in **Figure 20.** A **photovoltaic** is a device that transforms radiant energy directly into electrical energy. Photovoltaics are used to power calculators and satellites, including the *International Space Station.*

Figure 20 Solar energy can be collected and utilized by individuals using thermal collectors or photovoltaic collectors.

Reading Check *What does a photovoltaic do?*

Geothermal Energy

Imagine you could take a journey to the center of Earth—down to about 6,400 km below the surface. As you went deeper and deeper, you would find the temperature increasing. In fact, after going only about 3 km, the temperature could have increased enough to boil water. At a depth of 100 km, the temperature could be over 900°C. The heat generated inside Earth is called geothermal energy. Some of this heat is produced when unstable radioactive atoms inside Earth decay, converting nuclear energy to thermal energy.

At some places deep within Earth the temperature is hot enough to melt rock. This molten rock, or magma, can rise up close to the surface through cracks in the crust. During a volcanic eruption, magma reaches the surface. In other places, magma gets close to the surface and heats the rock around it.

Geothermal Reservoirs In some regions where magma is close to the surface, rainwater and water from melted snow can seep down to the hot rock through cracks and other openings in Earth's surface. The water then becomes hot and sometimes can form steam. The hot water and steam can be trapped under high pressure in cracks and pockets called geothermal reservoirs. In some places, the hot water and steam are close enough to the surface to form hot springs and geysers.

Geothermal Power Plants In places where the geothermal reservoirs are less than several kilometers deep, wells can be drilled to reach them. The hot water and steam produced by geothermal energy then can be used by geothermal power plants, like the one in **Figure 21,** to generate electricity.

Most geothermal reservoirs contain hot water under high pressure. **Figure 22** shows how these reservoirs can be used to generate electricity. While geothermal power is an inexhaustible source of energy, geothermal power plants can be built only in regions where geothermal reservoirs are close to the surface, such as in the western United States.

Heat Pumps Geothermal heat helps keep the temperature of the ground at a depth of several meters at a nearly constant temperature of about 10° to 20°C. This constant temperature can be used to cool and heat buildings by using a heat pump.

A heat pump contains a water-filled loop of pipe that is buried to a depth where the temperature is nearly constant. In summer the air is warmer than this underground temperature. Warm water from the building is pumped through the pipe down into the ground. The water cools and then is pumped back to the house where it absorbs more heat, and the cycle is repeated. During the winter, the air is cooler than the ground below. Then, cool water absorbs heat from the ground and releases it into the house.

Figure 21 This geothermal power plant in Nevada produces enough electricity to power about 50,000 homes.

Figure 22 The hot water in a geothermal reservoir is used to generate electricity in a geothermal power plant.

Energy from the Oceans

The ocean is in constant motion. If you've been to the seashore you've seen waves roll in. You may have seen the level of the ocean rise and fall over a period of about a half day. This rise and fall in the ocean level is called a tide. The constant movement of the ocean is an inexhaustible source of mechanical energy that can be converted into electric energy. While methods are still being developed to convert the motion in ocean waves to electric energy, several electric power plants using tidal motion have been built.

Figure 23 This tidal power plant in Annapolis Royal, Nova Scotia, is the only operating tidal power plant in North America.

Using Tidal Energy A high tide and a low tide each occur about twice a day. In most places the level of the ocean changes by less than a few meters. However, in some places the change is much greater. In the Bay of Fundy in Eastern Canada, the ocean level changes by 16 m between high tide and low tide. Almost 14 trillion kg of water move into or out of the bay between high and low tide.

Figure 23 shows an electric power plant that has been built along the Bay of Fundy. This power plant generates enough electric energy to power about 12,000 homes. The power plant is constructed so that as the tide rises, water flows through a turbine that causes an electric generator to spin, as shown in **Figure 24A.** The water is then trapped behind a dam. When the tide goes out, the trapped water behind the dam is released through the turbine to generate more electricity, as shown in **Figure 24B.** Each day electric power is generated for about ten hours when the tide is rising and falling.

While tidal energy is a nonpolluting, inexhaustible energy source, its use is limited. Only in a few places is the difference between high and low tide large enough to enable a large electric power plant to be built.

Figure 24 A tidal power plant can generate electricity when the tide is coming in and going out.

A As the tide comes in, it turns a turbine connected to a generator. When high tide occurs, gates are closed that trap water behind a dam.

B As the tide goes out and the ocean level drops, the gates are opened and water from behind the dam flows through the turbine, causing it to spin and turn a generator.

Wind

Wind is another inexhaustible supply of energy. Modern windmills, like the ones in **Figure 25,** convert the kinetic energy of the wind to electrical energy. The propeller is connected to a generator so that electrical energy is generated when wind spins the propeller. These windmills produce almost no pollution. Some disadvantages are that windmills produce noise and that large areas of land are needed. Also, studies have shown that birds sometimes are killed by windmills.

Conserving Energy

Fossil fuels are a valuable resource. Not only are they burned to provide energy, but oil and coal also are used to make plastics and other materials. One way to make the supply of fossil fuels last longer is to use less energy. Reducing the use of energy is called conserving energy.

You can conserve energy and also save money by turning off lights and appliances such as televisions when you are not using them. Also keep doors and windows closed tightly when it's cold or hot to keep heat from leaking out of or into your house. Energy could also be conserved if buildings are properly insulated, especially around windows. The use of oil could be reduced if cars were used less and made more efficient, so they went farther on a liter of gas. Recycling materials such as aluminum cans and glass also helps conserve energy.

Figure 25 Windmills work on the same basic principles as a power plant. Instead of steam turning a turbine, wind turns the rotors. **Describe** *some of the advantages and disadvantages of using windmills.*

section 3 review

Summary

Nonrenewable Resources

- All energy resources have advantages and disadvantages.
- Nonrenewable energy resources are used faster than they are replaced.
- Fossil fuels include oil, coal, and natural gas and are nonrenewable resources. Nuclear energy is a nonrenewable resource.

Renewable and Alternative Resources

- Renewable energy resources, such as hydroelectricity, are resources that are replenished continually.
- Alternative energy sources include solar energy, wind energy, and geothermal energy.

Self Check

1. **Diagram** the energy conversions that occur when coal is formed, and then burned to produce thermal energy.
2. **Explain** why solar energy is considered an inexhaustible source of energy.
3. **Explain** how a heat pump is used to both heat and cool a building.
4. **Think Critically** Identify advantages and disadvantages of using fossil fuels, hydroelectricity, and solar energy as energy sources.

Applying Math

5. **Use a Ratio** Earth's temperature increases with depth. Suppose the temperature increase inside Earth is 500°C at a depth of 50 km. What is the temperature increase at a depth of 10 km?

LAB Use the Internet

Energy to Power Your Life

Goals
- **Identify** how energy you use is produced and delivered.
- **Investigate** alternative sources for the energy you use.
- **Outline** a plan for how these alternative sources of energy could be used.

Data Source

Visit bookm.msscience.com/internet_lab for more information about sources of energy and for data collected by other students.

Real-World Question

Over the past 100 years, the amount of energy used in the United States and elsewhere has greatly increased. Today, a number of energy sources are available, such as coal, oil, natural gas, nuclear energy, hydroelectric power, wind, and solar energy. Some of these energy sources are being used up and are nonrenewable, but others are replaced as fast as they are used and, therefore, are renewable. Some energy sources are so vast that human usage has almost no effect on the amount available. These energy sources are inexhaustible.

Think about the types of energy you use at home and school every day. In this lab, you will investigate how and where energy is produced, and how it gets to you. You will also investigate alternative ways energy can be produced, and whether these sources are renewable, nonrenewable, or inexhaustible. What are the sources of the energy you use every day?

Local Energy Information	
Energy Type	
Where is that energy produced?	
How is that energy produced?	**Do not write in this book.**
How is that energy delivered to you?	
Is the energy source renewable, nonrenewable, or inexhaustible?	
What type of alternative energy source could you use instead?	

148 ◆ **M** CHAPTER 5 Energy and Energy Resources

Using Scientific Methods

▶ Make a Plan

1. Think about the activities you do every day and the things you use. When you watch television, listen to the radio, ride in a car, use a hair drier, or turn on the air conditioning, you use energy. Select one activity or appliance that uses energy.
2. **Identify** the type of energy that is used.
3. **Investigate** how that energy is produced and delivered to you.
4. **Determine** if the energy source is renewable, nonrenewable, or inexhaustible.
5. If your energy source is nonrenewable, describe how the energy you use could be produced by renewable sources.

▶ Follow Your Plan

1. Make sure your teacher approves your plan before you start.
2. Organize your findings in a data table, similar to the one that is shown.

▶ Analyze Your Data

1. **Describe** the process for producing and delivering the energy source you researched. How is it created, and how does it get to you?
2. How much energy is produced by the energy source you investigated?
3. Is the energy source you researched renewable, nonrenewable, or inexhaustible? Why?

▶ Conclude and Apply

1. **Describe** If the energy source you investigated is nonrenewable, how can the use of this energy source be reduced?
2. **Organize** What alternative sources of energy could you use for everyday energy needs? On the computer, create a plan for using renewable or inexhaustible sources.

Communicating Your Data

Find this lab using the link below. Post your data in the table that is provided. **Compare** your data to those of other students. **Combine** your data with those of other students and make inferences using the combined data.

Science online
bookm.msscience.com/internet_lab

LAB **M** ♦ 149

SCIENCE Stats

Energy to Burn

Did you know...

...The energy released by the average hurricane is equal to about 200 times the total energy produced by all of the world's power plants. Almost all of this energy is released as heat when raindrops form.

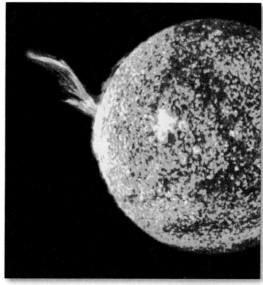

...The energy Earth gets each half hour from the Sun is enough to meet the world's demands for a year. Renewable and inexhaustible resources, including the Sun, account for only 18 percent of the energy that is used worldwide.

...The Calories in one medium apple will give you enough energy to walk for about 15 min, swim for about 10 min, or jog for about 9 min.

Applying Math If walking for 15 min requires 80 Calories of fuel (from food), how many Calories would someone need to consume to walk for 1 h?

Write About It

Where would you place solar collectors in the United States? Why? For more information on solar energy, go to bookm.msscience.com/science_stats.

Chapter 5 Study Guide

Reviewing Main Ideas

Section 1 — What is energy?

1. Energy is the ability to cause change.
2. A moving object has kinetic energy that depends on the object's mass and speed.
3. Potential energy is energy due to position and depends on an object's mass and height.
4. Light carries radiant energy, electric current carries electrical energy, and atomic nuclei contain nuclear energy.

Section 2 — Energy Transformations

1. Energy can be transformed from one form to another. Thermal energy is usually produced when energy transformations occur.
2. The law of conservation of energy states that energy cannot be created or destroyed.

3. Electric power plants convert a source of energy into electrical energy. Steam spins a turbine which spins an electric generator.

Section 3 — Sources of Energy

1. The use of an energy source has advantages and disadvantages.
2. Fossil fuels and nuclear energy are nonrenewable energy sources that are consumed faster than they can be replaced.
3. Hydroelectricity is a renewable energy source that is continually being replaced.
4. Alternative energy sources include solar, wind, and geothermal energy. Solar energy is an inexhaustible energy source.

Visualizing Main Ideas

Copy and complete the concept map using the following terms: fossil fuels, hydroelectric, solar, wind, oil, coal, photovoltaic, *and* nonrenewable resources.

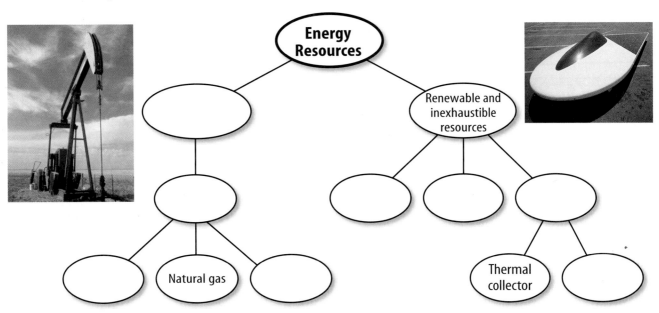

chapter 5 Review

Using Vocabulary

alternative resource p. 143
chemical energy p. 129
electrical energy p. 130
energy p. 126
generator p. 136
inexhaustible resource p. 143
kinetic energy p. 127
law of conservation of energy p. 132
nonrenewable resource p. 140
nuclear energy p. 130
photovoltaic p. 144
potential energy p. 128
radiant energy p. 129
renewable resource p. 142
thermal energy p. 128
turbine p. 136

For each of the terms below, explain the relationship that exists.

1. electrical energy—nuclear energy
2. turbine—generator
3. photovoltaic—radiant energy—electrical energy
4. renewable resource—inexhaustible resource
5. potential energy—kinetic energy
6. kinetic energy—electrical energy—generator
7. thermal energy—radiant energy
8. law of conservation of energy—energy transformations
9. nonrenewable resource—chemical energy

Checking Concepts

Choose the word or phrase that best answers the question.

10. Objects that are able to fall have what type of energy?
    A) kinetic C) potential
    B) radiant D) electrical

11. Which form of energy does light have?
    A) electrical C) kinetic
    B) nuclear   D) radiant

12. Muscles perform what type of energy transformation?
    A) kinetic to potential
    B) kinetic to electrical
    C) thermal to radiant
    D) chemical to kinetic

13. Photovoltaics perform what type of energy transformation?
    A) thermal to radiant
    B) kinetic to electrical
    C) radiant to electrical
    D) electrical to thermal

14. The form of energy that food contains is which of the following?
    A) chemical   C) radiant
    B) potential D) electrical

15. Solar energy, wind, and geothermal are what type of energy resource?
    A) inexhaustible C) nonrenewable
    B) inexpensive   D) chemical

16. Which of the following is a nonrenewable source of energy?
    A) hydroelectricity
    B) nuclear
    C) wind
    D) solar

17. A generator is NOT required to generate electrical energy when which of the following energy sources is used?
    A) solar C) hydroelectric
    B) wind   D) nuclear

18. Which of the following are fossil fuels?
    A) gas   C) oil
    B) coal D) all of these

19. Almost all of the energy that is used on Earth's surface comes from which of the following energy sources?
    A) radioactivity C) chemicals
    B) the Sun   D) wind

152 ◆ M CHAPTER REVIEW

chapter 5 Review

Thinking Critically

20. **Explain** how the motion of a swing illustrates the transformation between potential and kinetic energy.

21. **Explain** what happens to the kinetic energy of a skateboard that is coasting along a flat surface, slows down, and comes to a stop.

22. **Describe** the energy transformations that occur in the process of toasting a bagel in an electric toaster.

23. **Compare and contrast** the formation of coal and the formation of oil and natural gas.

24. **Explain** the difference between the law of conservation of energy and conserving energy. How can conserving energy help prevent energy shortages?

25. **Make a Hypothesis** about how spacecraft that travel through the solar system obtain the energy they need to operate. Do research to verify your hypothesis.

26. **Concept Map** Copy and complete this concept map about energy.

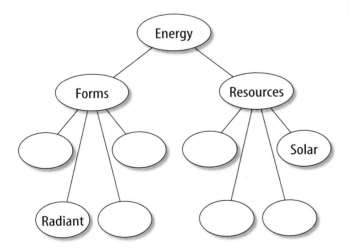

27. **Diagram** the energy transformations that occur when you rub sandpaper on a piece of wood and the wood becomes warm.

Performance Activities

28. **Multimedia Presentation** Alternative sources of energy that weren't discussed include biomass energy, wave energy, and hydrogen fuel cells. Research an alternative energy source and then prepare a digital slide show about the information you found. Use the concepts you learned from this chapter to inform your classmates about the future prospects of using such an energy source on a large scale.

Applying Math

29. **Calculate Number of Power Plants** A certain type of power plant is designed to provide energy for 10,000 homes. How many of these power plants would be needed to provide energy for 300,000 homes?

Use the table below to answer questions 30 and 31.

Energy Sources Used in the United States	
Energy Source	Percent of Energy Used
Coal	23%
Oil	39%
Natural gas	23%
Nuclear	8%
Hydroelectric	4%
Other	3%

30. **Use Percentages** According to the data in the table above, what percentage of the energy used in the United States comes from fossil fuels?

31. **Calculate a Ratio** How many times greater is the amount of energy that comes from fossil fuels than the amount of energy from all other energy sources?

Part 1 | Multiple Choice

Record your answers on the answer sheet provided by your teacher or on a sheet of paper.

1. The kinetic energy of a moving object increases if which of the following occurs?
   A. Its mass decreases.
   B. Its speed increases.
   C. Its height above the ground increases.
   D. Its temperature increases.

Use the graph below to answer questions 2–4.

2. According to the graph above, in which year will global oil production be at a maximum?
   A. 1974
   B. 2002
   C. 2010
   D. 2050

3. Approximately how many times greater was oil production in 1970 than oil production in 1950?
   A. 2 times
   B. 10 times
   C. 6 times
   D. 3 times

4. In which year will the production of oil be equal to the oil production in 1970?
   A. 2010
   B. 2015
   C. 2022
   D. 2028

5. Which of the following energy sources is being used faster than it can be replaced?
   A. tidal
   B. wind
   C. fossil fuels
   D. hydroelectric

Use the circle graph below to answer question 6.

6. The circle graph shows the sources of electrical energy in the United States. In 2002, the total amount of electrical energy produced in the United States was 38.2 quads. How much electrical energy was produced by nuclear power plants?
   A. 3.0 quads
   B. 3.8 quads
   C. 7.6 quads
   D. 35.1 quads

7. When chemical energy is converted into thermal energy, which of the following must be true?
   A. The total amount of thermal energy plus chemical energy changes.
   B. Only the amount of chemical energy changes.
   C. Only the amount of thermal energy changes.
   D. The total amount of thermal energy plus chemical energy doesn't change.

8. A softball player hits a fly ball. Which of the following describes the energy conversion that occurs as it falls from its highest point?
   A. kinetic to potential
   B. potential to kinetic
   C. thermal to potential
   D. thermal to kinetic

Standardized Test Practice

Part 2 Short Response/Grid In

Record your answers on the answer sheet provided by your teacher or on a sheet of paper.

9. Why is it impossible to build a machine that produces more energy than it uses?

10. You toss a ball upward and then catch it on the way down. The height of the ball above the ground when it leaves your hand on the way up and when you catch it is the same. Compare the ball's kinetic energy when it leaves your hand and just before you catch it.

11. A basket ball is dropped from a height of 2 m and another identical basketball is dropped from a height of 4 m. Which ball has more kinetic energy just before it hits the ground?

Use the graph below to answer questions 12 and 13.

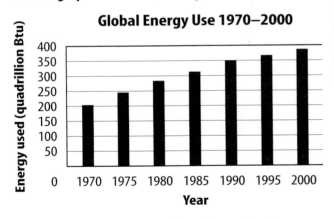

12. According to the graph above, by about how many times did the global use of energy increase from 1970 to 2000?

13. Over which five-year time period was the increase in global energy use the largest?

Test-Taking Tip

Do Your Studying Regularly Do not "cram" the night before the test. It can hamper your memory and make you tired.

Part 3 Open Ended

Record your answers on a sheet of paper.

14. When you drop a tennis ball, it hits the floor and bounces back up. But it does not reach the same height as released, and each successive upward bounce is smaller than the one previous. However, you notice the tennis ball is slightly warmer after it finishes bouncing. Explain how the law of conservation of energy is obeyed.

Use the graph below to answer questions 15–17.

15. The graph shows how the potential energy of a batted ball depends on distance from the batter. At what distances is the kinetic energy of the ball the greatest?

16. At what distance from the batter is the height of the ball the greatest?

17. How much less is the kinetic energy of the ball at a distance of 20 m from the batter than at a distance of 0 m?

18. List advantages and disadvantages of the following energy sources: fossil fuels, nuclear energy, and geothermal energy.

chapter 6

Thermal Energy

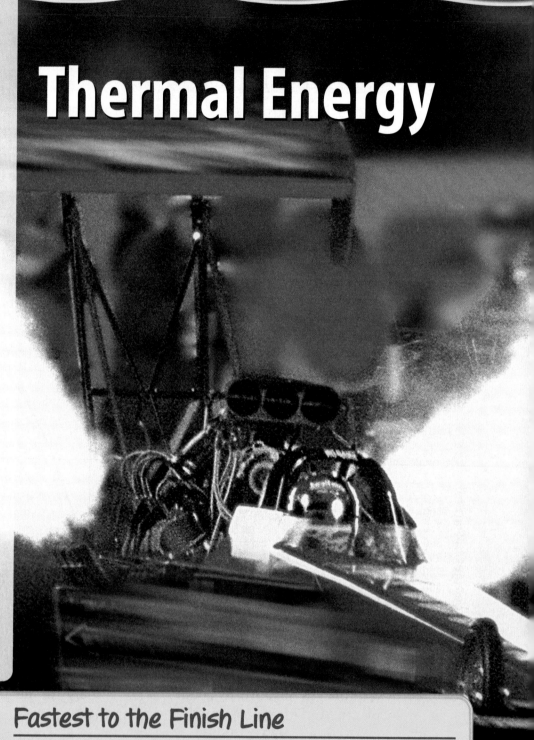

The BIG Idea
Thermal energy flows from areas of higher temperature to areas of lower temperature.

SECTION 1
Temperature and Thermal Energy
Main Idea Atoms and molecules in an object are moving in all directions with different speeds.

SECTION 2
Heat
Main Idea Thermal energy moves by conduction, convection, and radiation.

SECTION 3
Engines and Refrigerators
Main Idea Engines convert thermal energy to mechanical energy; refrigerators transfer thermal energy from one place to another.

Fastest to the Finish Line
In order to reach an extraordinary speed in a short distance, this dragster depends on more than an aerodynamic design. Its engine must transform the thermal energy produced by burning fuel to mechanical energy, which propels the dragster down the track.

Science Journal Describe five things that you do to make yourself feel warmer or cooler.

Start-Up Activities

Measuring Temperature

When you leave a glass of ice water on a kitchen table, the ice gradually melts and the temperature of the water increases. What is temperature, and why does the temperature of the ice water increase? In this lab you will explore one way of determining temperature.

1. Obtain three pans. Fill one pan with lukewarm water. Fill a second pan with cold water and crushed ice. Fill a third pan with very warm tap water. Label each pan.
2. Soak one of your hands in the warm water for one minute. Remove your hand from the warm water and put it in the lukewarm water. Does the lukewarm water feel cool or warm?
3. Now soak your hand in the cold water for one minute. Remove your hand from the cold water and place it in the lukewarm water. Does the lukewarm water feel cool or warm?
4. **Think Critically** Write a paragraph in your Science Journal discussing whether your sense of touch would make a useful thermometer.

Thermal Energy Make the following Foldable to help you identify how thermal energy, heat, and temperature are related.

STEP 1 Fold a vertical piece of paper into thirds.

STEP 2 Turn the paper horizontally. Unfold and label the three columns as shown.

Read for Main Ideas Before you read the chapter, write down what you know about temperature, thermal energy, and heat on the appropriate tab. As you read, add to and correct what you wrote. Write what you have learned about the relationship between heat and thermal energy on the back of your Foldable.

 Preview this chapter's content and activities at bookm.msscience.com

Get Ready to Read

Identify the Main Idea

① Learn It! Main ideas are the most important ideas in a paragraph, a section, or a chapter. Supporting details are facts or examples that explain the main idea. Understanding the main idea allows you to grasp the whole picture.

② Picture It! Read the following paragraph. Draw a graphic organizer like the one below to show the main idea and supporting details.

> When you heat a pot of water on a stove, thermal energy can be transferred through the water by a process other than conduction and radiation. In a gas or liquid, atoms or molecules can move much more easily than they can in a solid. As a result, these particles can travel from one place to another, carrying their energy with them. This transfer of thermal energy by the movement of atoms or molecules from one part of a material to another is called convection.
>
> —from page 164

```
                    Main Idea
                   /    |    \
      Supporting  Supporting  Supporting
       Details    Details     Details
```

③ Apply It! Pick a paragraph from another section of this chapter and diagram the main ideas as you did above.

Target Your Reading

Reading Tip: The main idea is often the first sentence in a paragraph, but not always.

Use this to focus on the main ideas as you read the chapter.

1 Before you read the chapter, respond to the statements below on your worksheet or on a numbered sheet of paper.
- Write an **A** if you **agree** with the statement.
- Write a **D** if you **disagree** with the statement.

2 After you read the chapter, look back to this page to see if you've changed your mind about any of the statements.
- If any of your answers changed, explain why.
- Change any false statements into true statements.
- Use your revised statements as a study guide.

Science Online
Print out a worksheet of this page at bookm.msscience.com

Before You Read A or D		Statement	After You Read A or D
	1	Temperature depends on the kinetic energy of the molecules in a material.	
	2	Heat engines can convert energy from one form to another.	
	3	Objects cannot have a temperature below zero on the Celsius scale.	
	4	In a refrigerator, the coolant gas gets cooler as it is compressed.	
	5	A conductor is any material that easily transfers thermal energy.	
	6	Energy is created by an engine.	
	7	Thermal energy from the Sun reaches Earth by conduction through space.	
	8	A car's engine converts thermal energy to mechanical energy.	
	9	Thermal energy always moves from colder objects to warmer objects.	

M ◆ 158 B

section 1
Temperature and Thermal Energy

as you read

What You'll Learn
- **Explain** how temperature is related to kinetic energy.
- **Describe** three scales used for measuring temperature.
- **Define** thermal energy.

Why It's Important
The movement of thermal energy toward or away from your body determines whether you feel too cold, too hot, or just right.

Review Vocabulary
kinetic energy: energy a moving object has that increases as the speed of the object increases

New Vocabulary
- temperature
- thermal energy

What is temperature?

Imagine it's a hot day and you jump into a swimming pool to cool off. When you first hit the water, you might think it feels cold. Perhaps someone else, who has been swimming for a few minutes, thinks the water feels warm. When you swim in water, touch a hot pan, or swallow a cold drink, your sense of touch tells you whether something is hot or cold. However, the words *cold*, *warm*, and *hot* can mean different things to different people.

Temperature How hot or cold something feels is related to its temperature. To understand temperature, think of a glass of water sitting on a table. The water might seem perfectly still, but water is made of molecules that are in constant, random motion. Because these molecules are always moving, they have energy of motion, or kinetic energy.

However, water molecules in random motion don't all move at the same speed. Some are moving faster and some are moving slower. **Temperature** is a measure of the average value of the kinetic energy of the molecules in random motion. The more kinetic energy the molecules have, the higher the temperature. Molecules have more kinetic energy when they are moving faster. So the higher the temperature, the faster the molecules are moving, as shown in **Figure 1**.

Figure 1 The temperature of a substance depends on how fast its molecules are moving. Water molecules are moving faster in the hot water on the left than in the cold water on the right.

Hot water

Cold water

Thermal Expansion It wasn't an earthquake that caused the sidewalk to buckle in **Figure 2.** Hot weather caused the concrete to expand so much that it cracked, and the pieces squeezed each other upward. When the temperature of an object is increased, its molecules speed up and tend to move farther apart. This causes the object to expand. When the object is cooled, its molecules slow down and move closer together. This causes the object to shrink, or contract.

Almost all substances expand when they are heated and contract when they are cooled. The amount of expansion or contraction depends on the type of material and the change in temperature. For example, liquids usually expand more than solids. Also, the greater the change in temperature, the more an object expands or contracts.

 Why do materials expand when their temperatures increase?

Figure 2 Most objects expand as their temperatures increase. Pieces of this concrete sidewalk forced each other upward when the concrete expanded on a hot day.

Measuring Temperature

The temperature of an object depends on the average kinetic energy of all the molecules in an object. However, molecules are so small and objects contain so many of them, that it is impossible to measure the kinetic energy of all the individual molecules.

A more practical way to measure temperature is to use a thermometer. Thermometers usually use the expansion and contraction of materials to measure temperature. One common type of thermometer uses a glass tube containing a liquid. When the temperature of the liquid increases, it expands so that the height of the liquid in the tube depends on the temperature.

Temperature Scales To be able to give a number for the temperature, a thermometer has to have a temperature scale. Two common temperature scales are the Fahrenheit and Celsius scales, shown in **Figure 3.**

On the Fahrenheit scale, the freezing point of water is given the temperature 32°F and the boiling point 212°F. The space between the boiling point and the freezing point is divided into 180 equal degrees. The Fahrenheit scale is used mainly in the United States.

On the Celsius temperature scale, the freezing point of water is given the temperature 0°C and the boiling point is given the temperature 100°C. Because there are only 100 Celsius degrees between the boiling and freezing point of water, Celsius degrees are bigger than Fahrenheit degrees.

Figure 3 The Fahrenheit and Celsius scales are commonly used temperature scales.

Converting Fahrenheit and Celsius You can convert temperatures back and forth between the two temperature scales by using the following equations.

Temperature Conversion Equations

To convert temperature in °F to °C: $°C = (\frac{5}{9})(°F - 32)$

To convert temperature in °C to °F: $°F = (\frac{9}{5})(°C) + 32$

For example, to convert 68°F to degrees Celsius, first subtract 32, multiply by 5, then divide by 9. The result is 20°C.

The Kelvin Scale Another temperature scale that is sometimes used is the Kelvin scale. On this scale, 0 K is the lowest temperature an object can have. This temperature is known as absolute zero. The size of a degree on the Kelvin scale is the same as on the Celsius scale. You can change from Celsius degrees to Kelvin degrees by adding 273 to the Celsius temperature.

$$K = °C + 273$$

Applying Math — Solving a Simple Equation

CONVERTING TO CELSIUS On a hot summer day, a Fahrenheit thermometer shows the temperature to be 86°F. What is this temperature on the Celsius scale?

Solution

① *This is what you know:* Fahrenheit temperature: °F = 86

② *This is what you need to find:* Celsius temperature: °C

③ *This is the procedure you need to use:* Substitute the Fahrenheit temperature into the equation that converts temperature in °F to °C.
$°C = (\frac{5}{9})(°F - 32) = \frac{5}{9}(86 - 32) = \frac{5}{9}(54) = 30°C$

④ *Check the answer:* Add 32 to your answer and multiply by 9/5. The result should be the given Fahrenheit temperature.

Practice Problems

1. A student's body temperature is 98.6°F. What is this temperature on the Celsius scale?
2. A temperature of 57°C was recorded in 1913 at Death Valley, California. What is this temperature on the Fahrenheit scale?

For more practice visit
bookm.msscience.com/
math_practice

Thermal Energy

The temperature of an object is related to the average kinetic energy of molecules in random motion. But molecules also have potential energy. Potential energy is energy that the molecules have that can be converted into kinetic energy. The sum of the kinetic and potential energy of all the molecules in an object is the **thermal energy** of the object.

The Potential Energy of Molecules When you hold a ball above the ground, it has potential energy. When you drop the ball, its potential energy is converted into kinetic energy as the ball falls toward Earth. It is the attractive force of gravity between Earth and the ball that gives the ball potential energy.

The molecules in a material also exert attractive forces on each other. As a result, the molecules in a material have potential energy. As the molecules get closer together or farther apart, their potential energy changes.

Increasing Thermal Energy Temperature and thermal energy are different. Suppose you have two glasses filled with the same amount of milk, and at the same temperature. If you pour both glasses of milk into a pitcher, as shown in **Figure 4,** the temperature of the milk won't change. However, because there are more molecules of milk in the pitcher than in either glass, the thermal energy of the milk in the pitcher is greater than the thermal energy of the milk in either glass.

Figure 4 At the same temperature, the larger volume of milk in the pitcher has more thermal energy than the smaller volumes of milk in either glass.

section 1 review

Summary

Temperature
- Temperature is related to the average kinetic energy of the molecules an object contains.
- Most materials expand when their temperatures increase.

Measuring Temperature
- On the Celsius scale the freezing point of water is 0°C and the boiling point is 100°C.
- On the Fahrenheit scale the freezing point of water is 32°F and the boiling point is 212°F.

Thermal Energy
- The thermal energy of an object is the sum of the kinetic and potential energy of all the molecules in an object.

Self Check

1. **Explain** the difference between temperature and thermal energy. How are they related?
2. **Determine** which temperature is always larger—an object's Celsius temperature or its Kelvin temperature.
3. **Explain** how kinetic energy and thermal energy are related.
4. **Describe** how a thermometer uses the thermal expansion of a material to measure temperature.

Applying Math

5. **Convert Temperatures** A turkey cooking in an oven will be ready when the internal temperature reaches 180°F. Convert this temperature to °C and K.

section 2
Heat

as you read

What You'll Learn
- **Explain** the difference between thermal energy and heat.
- **Describe** three ways thermal energy is transferred.
- **Identify** materials that are insulators or conductors.

Why It's Important
To keep you comfortable, the flow of thermal energy into and out of your house must be controlled.

Review Vocabulary
electromagnetic wave: a wave produced by vibrating electric charges that can travel in matter and empty space

New Vocabulary
- heat
- conduction
- radiation
- convection
- conductor
- specific heat
- thermal pollution

Heat and Thermal Energy

It's the heat of the day. Heat the oven to 375°F. A heat wave has hit the Midwest. You've often heard the word *heat*, but what is it? Is it something you can see? **Heat** is the transfer of thermal energy from one object to another when the objects are at different temperatures. The amount of thermal energy that is transferred when two objects are brought into contact depends on the difference in temperature between the objects.

For example, no thermal energy is transferred when two pots of boiling water are touching, because the water in both pots is at the same temperature. However, thermal energy is transferred from the pot of hot water in **Figure 5** that is touching a pot of cold water. The hot water cools down and the cold water gets hotter. Thermal energy continues to be transferred until both objects are the same temperature.

Transfer of Thermal Energy When thermal energy is transferred, it always moves from warmer to cooler objects. Thermal energy never flows from a cooler object to a warmer object. The warmer object loses thermal energy and becomes cooler as the cooler object gains thermal energy and becomes warmer. This process of thermal energy transfer can occur in three ways—by conduction, radiation, or convection.

Figure 5 Thermal energy is transferred only when two objects are at different temperatures. Thermal energy always moves from the warmer object to the cooler object.

Conduction

When you eat hot pizza, you experience conduction. As the hot pizza touches your mouth, thermal energy moves from the pizza to your mouth. This transfer of thermal energy by direct contact is called conduction. **Conduction** occurs when the particles in a material collide with neighboring particles.

Imagine holding an ice cube in your hand, as in **Figure 6**. The faster-moving molecules in your warm hand bump against the slower-moving molecules in the cold ice. In these collisions, energy is passed from molecule to molecule. Thermal energy flows from your warmer hand to the colder ice, and the slow-moving molecules in the ice move faster. As a result, the ice becomes warmer and its temperature increases. Molecules in your hand move more slowly as they lose thermal energy, and your hand becomes cooler.

Conduction usually occurs most easily in solids and liquids, where atoms and molecules are close together. Then atoms and molecules need to move only a short distance before they bump into one another. As a result, thermal energy is transferred more rapidly by conduction in solids and liquids than in gases.

 Why does conduction occur more easily in solids and liquids than in gases?

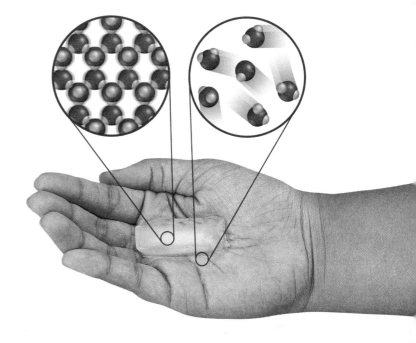

Figure 6 An ice cube in your hand melts because of conduction.

Radiation

On a clear day, you walk outside and feel the warmth of the Sun. How does this transfer of thermal energy occur? Thermal energy transfer does not occur by conduction because almost no matter exists between the Sun and Earth. Instead, thermal energy is transferred from the Sun to Earth by radiation. Thermal energy transfer by **radiation** occurs when energy is transferred by electromagnetic waves. These waves carry energy through empty space, as well as through matter. The transfer of thermal energy by radiation can occur in empty space, as well as in solids, liquids, and gases.

The Sun is not the only source of radiation. All objects emit electromagnetic radiation, although warm objects emit more radiation than cool objects. The warmth you feel when you sit next to a fireplace is due to the thermal energy transferred by radiation from the fire to your skin.

Comparing Rates of Melting

Procedure
1. Prepare ice water by filling a **glass** with ice and then adding water. Let the glass sit until all the ice melts.
2. Place an ice cube in a **coffee cup**.
3. Place a similar-sized ice cube in another **coffee cup** and add ice water to a depth of about 1 cm.
4. Time how long it takes both ice cubes to melt.

Analysis
1. Which ice cube melted fastest? Why?
2. Is air or water a better insulator? Explain.

Convection

When you heat a pot of water on a stove, thermal energy can be transferred through the water by a process other than conduction and radiation. In a gas or liquid, atoms or molecules can move much more easily than they can in a solid. As a result, these particles can travel from one place to another, carrying their energy along with them. This transfer of thermal energy by the movement of atoms or molecules from one part of a material to another is called **convection**.

Transferring Thermal Energy by Convection As a pot of water is heated, thermal energy is transferred by convection. First, thermal energy is transferred to the water molecules at the bottom of the pot from the stove. These water molecules move faster as their thermal energy increases. The faster-moving molecules tend to be farther apart than the slower-moving molecules in the cooler water above. This causes the warm water to be less dense than the cooler water. As a result, the warm water rises and is replaced at the bottom of the pot by cooler water. The cooler water is heated, rises, and the cycle is repeated until all the water in the pan is at the same temperature.

Natural Convection Natural convection occurs when a warmer, less dense fluid is pushed away by a cooler, denser fluid. For example, imagine the shore of a lake. During the day, the water is cooler than the land. As shown in **Figure 7**, air above the warm land is heated by conduction. When the air gets hotter, its particles move faster and get farther from each other, making the air less dense. The cooler, denser air from over the lake flows in over the land, pushing the less dense air upward. You feel this movement of incoming cool air as wind. The cooler air then is heated by the land and also begins to rise.

Figure 7 Wind movement near a lake or ocean can result from natural convection.

Figure 8 This computer uses forced convection to keep the electronic components surrounded by cooler air.
Identify another example of forced convection.

Forced Convection Sometimes convection can be forced. Forced convection occurs when an outside force pushes a fluid, such as air or water, to make it move and transfer thermal energy. A fan is one type of device that is used to move air. For example, computers use fans to keep their electronic components from getting too hot, which can damage them. The fan blows cool air onto the hot electronic components, as shown in **Figure 8.** Thermal energy from the electronic components is transferred to the air around them by conduction. The warm air is pushed away as cool air rushes in. The hot components then continue to lose thermal energy as the fan blows cool air over them.

Thermal Conductors

Why are cooking pans usually made of metal? Why does the handle of a metal spoon in a bowl of hot soup become warm? The answer to both questions is that metal is a good conductor. A **conductor** is any material that easily transfers thermal energy. Some materials are good conductors because of the types of atoms or chemical compounds they contain.

Reading Check *What is a conductor?*

Remember that an atom has a nucleus surrounded by one or more electrons. Certain materials, such as metals, have some electrons that are not held tightly by the nucleus and are freer to move around. These loosely held electrons can bump into other atoms and help transfer thermal energy. The best conductors of thermal energy are metals such as gold and copper.

Mini LAB

Observing Convection

Procedure

1. Fill a **250-mL beaker** with room-temperature **water** and let it stand undisturbed for at least 1 min.
2. Using a **hot plate,** heat a small amount of water in a **50-mL beaker** until it is almost boiling.
   WARNING: *Do not touch the heated hot plate.*
3. Carefully place a **penny** into the hot water and let it stand for about 1 min.
4. Take the penny out of the hot water with **metal tongs** and place it on a table. Immediately place the 250-mL beaker on the penny.
5. Using a **dropper,** gently place one drop of **food coloring** on the bottom of the 250-mL beaker of water.
6. Observe what happens in the beaker for several minutes.

Analysis
What happened when you placed the food coloring in the 250-mL beaker? Why?

Animal Insulation
To survive in its arctic environment, a polar bear needs good insulation against the cold. Underneath its fur, a polar bear has 10 cm of insulating blubber. Research how animals in polar regions are able to keep themselves warm. Summarize the different ways in your Science Journal.

Thermal Insulators

If you're cooking food, you want the pan to conduct thermal energy easily from the stove to your food, but you do not want the handle of the pan to become hot. An insulator is a material in which thermal energy doesn't flow easily. Most pans have handles that are made from insulators. Liquids and gases are usually better insulators than solids are. Air is a good insulator, and many insulating materials contain air spaces that reduce the transfer of thermal energy by conduction within the material. Materials that are good conductors, such as metals, are poor insulators, and poor conductors are good insulators.

Houses and buildings are made with insulating materials to reduce the flow of thermal energy between the inside and outside. Fluffy insulation like that shown in **Figure 9** is put in the walls. Some windows have double layers of glass that sandwich a layer of air or other insulating gas. This reduces the outward flow of thermal energy in the winter and the inward flow of thermal energy in the summer.

Heat Absorption

On a hot day, you can walk barefoot across the lawn, but the asphalt pavement of a street is too hot to walk on. Why is the pavement hotter than the grass? The change in temperature of an object as it is heated depends on the material it is made of.

Specific Heat The temperature change of a material as it is heated depends on the specific heat of the material. The **specific heat** of a material is the amount of thermal energy needed to raise the temperature of 1 kg of the material by 1°C.

More thermal energy is needed to change the temperature of a material with a high specific heat than one with a low specific heat. For example, sand on a beach has a lower specific heat than water. During the day, radiation from the Sun warms the sand and the water. Because of its lower specific heat, the sand heats up faster than the water. At night, however, the sand feels cool and the water feels warmer. The temperature of the water changes more slowly than the temperature of the sand as they both lose thermal energy to the cooler night air.

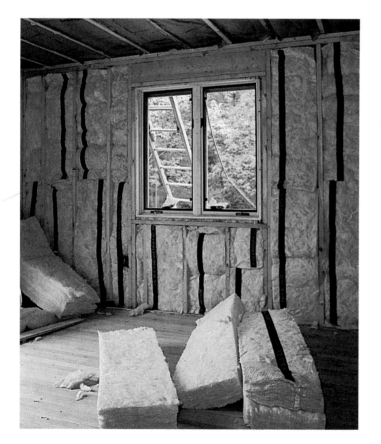

Figure 9 The insulation in houses and buildings helps reduce the transfer of thermal energy between the air inside and air outside.

Thermal Pollution

 Some electric power plants and factories that use water for cooling produce hot water as a by-product. If this hot water is released into an ocean, lake, or river, it will raise the temperature of the water nearby. This increase in the temperature of a body of water caused by adding warmer water is called **thermal pollution.** Rainwater that is heated after it falls on warm roads or parking lots also can cause thermal pollution if it runs off into a river or lake.

Effects of Thermal Pollution Increasing the water temperature causes fish and other aquatic organisms to use more oxygen. Because warmer water contains less dissolved oxygen than cooler water, some organisms can die due to a lack of oxygen. Also, in warmer water, many organisms become more sensitive to chemical pollutants, parasites, and diseases.

Reducing Thermal Pollution Thermal pollution can be reduced by cooling the warm water produced by factories, power plants, and runoff before it is released into a body of water. Cooling towers like the ones shown in **Figure 10** are used to cool the water used by some power plants and factories.

Figure 10 This power plant uses cooling towers to cool the warm water produced by the power plant.

section 2 review

Summary

Heat and Thermal Energy
- Heat is the transfer of thermal energy due to a temperature difference.
- Thermal energy always moves from a higher temperature to a lower temperature.

Conduction, Radiation, and Convection
- Conduction is the transfer of thermal energy when substances are in direct contact.
- Radiation is the transfer of thermal energy by electromagnetic waves.
- Convection is the transfer of thermal energy by the movement of matter.

Thermal Conductors and Specific Heat
- A thermal conductor is a material in which thermal energy moves easily.
- The specific heat of a substance is the amount of thermal energy needed to raise the temperature of 1 kg of the substance by 1°C.

Self Check

1. **Explain** why materials such as plastic foam, feathers, and fur are poor conductors of thermal energy.
2. **Explain** why the sand on a beach cools down at night more quickly than the ocean water.
3. **Infer** If a substance can contain thermal energy, can a substance also contain heat?
4. **Describe** how thermal energy is transferred from one place to another by convection.
5. **Explain** why a blanket keeps you warm.
6. **Think Critically** In order to heat a room evenly, should heating vents be placed near the floor or near the ceiling of the room? Explain.

Applying Skills

7. **Design an Experiment** to determine whether wood or iron is a better thermal conductor. Identify the dependent and independent variables in your experiment.

Heating Up and Cooling Down

Do you remember how long it took for a cup of hot chocolate to cool before you could take a sip? The hotter the chocolate, the longer it seemed to take to cool.

Real-World Question

How does the temperature of a liquid affect how quickly it warms or cools?

Goals
- Measure the temperature change of water at different temperatures.
- Infer how the rate of heating or cooling depends on the initial water temperature.

Materials
thermometers (5)
400-mL beakers (5)
stopwatch
*watch with second hand
hot plate
*Alternate materials

Safety Precautions

WARNING: *Do not use mercury thermometers. Use caution when heating with a hot plate. Hot and cold glass appears the same.*

Procedure

1. Make a data table to record the temperature of water in five beakers every minute from 0 to 10 min.
2. Fill one beaker with 100 mL of water. Place the beaker on a hot plate and bring the water to a boil. Carefully remove the hot beaker from the hot plate.

3. Record the water temperature at minute 0, and then every minute for 10 min.
4. Repeat step 3 starting with hot tap water, cold tap water, refrigerated water, and ice water with the ice removed.

Conclude and Apply

1. **Graph** your data. **Plot and label** lines for all five beakers on one graph.
2. **Calculate** the rate of heating or cooling for the water in each beaker by subtracting the initial temperature of the water from the final temperature and then dividing this answer by 10 min.
3. **Infer** from your results how the difference between room temperature and the initial temperature of the water affected the rate at which it heated up or cooled down.

Communicating Your Data

Share your data and graphs with other classmates and explain any differences among your data.

section 3
Engines and Refrigerators

Heat Engines

The engines used in cars, motorcycles, trucks, and other vehicles, like the one shown in **Figure 11,** are heat engines. A **heat engine** is a device that converts thermal energy into mechanical energy. Mechanical energy is the sum of the kinetic and potential energy of an object. The heat engine in a car converts thermal energy into mechanical energy when it makes the car move faster, causing the car's kinetic energy to increase.

Forms of Energy There are other forms of energy besides thermal energy and mechanical energy. For example, chemical energy is energy stored in the chemical bonds between atoms. Radiant energy is the energy carried by electromagnetic waves. Nuclear energy is energy stored in the nuclei of atoms. Electrical energy is the energy carried by electric charges as they move in a circuit. Devices such as heat engines convert one form of energy into other useful forms.

The Law of Conservation of Energy When energy is transformed from one form to another, the total amount of energy doesn't change. According to the law of conservation of energy, energy cannot be created or destroyed. Energy only can be transformed from one form to another. No device, including a heat engine, can produce energy or destroy energy.

as you read

What You'll Learn
- **Describe** what a heat engine does.
- **Explain** that energy can exist in different forms, but is never created or destroyed.
- **Describe** how an internal combustion engine works.
- **Explain** how refrigerators move thermal energy.

Why It's Important
Heat engines enable you to travel long distances.

Review Vocabulary
work: a way of transferring energy by exerting a force over a distance

New Vocabulary
- heat engine
- internal combustion engine

Figure 11 The engine in this earth mover transforms thermal energy into mechanical energy that can perform useful work.

Figure 12 Internal combustion engines are found in many tools and machines.

Internal Combustion Engines The heat engine you are probably most familiar with is the internal combustion engine. In **internal combustion engines,** the fuel burns in a combustion chamber inside the engine. Many machines, including cars, airplanes, buses, boats, trucks, and lawn mowers, use internal combustion engines, as shown in **Figure 12.**

Most cars have an engine with four or more combustion chambers, or cylinders. Usually the more cylinders an engine has, the more power it can produce. Each cylinder contains a piston that can move up and down. A mixture of fuel and air is injected into a combustion chamber and ignited by a spark. When the fuel mixture is ignited, it burns explosively and pushes the piston down. The up-and-down motion of the pistons turns a rod called a crankshaft, which turns the wheels of the car. **Figure 13** shows how an internal combustion engine converts thermal energy to mechanical energy in a process called the four-stroke cycle.

Several kinds of internal combustion engines have been designed. In diesel engines, the air in the cylinder is compressed to such a high pressure that the highly flammable fuel ignites without the need for a spark plug. Many lawn mowers use a two-stroke gasoline engine. The first stroke is a combination of intake and compression. The second stroke is a combination of power and exhaust.

 How does the burning of fuel mixture cause a piston to move?

Topic: Automobile Engines
Visit bookm.msscience.com for Web links to information on how internal combustion engines were developed for use in cars.

Activity Make a time line showing the five important events in the development of the automobile engine.

NATIONAL GEOGRAPHIC VISUALIZING THE FOUR-STROKE CYCLE

Figure 13

Most modern cars are powered by fuel-injected internal combustion engines that have a four-stroke combustion cycle. Inside the engine, thermal energy is converted into mechanical energy as gasoline is burned under pressure inside chambers known as cylinders. The steps in the four-stroke cycle are shown here.

EXHAUST STROKE

D The exhaust valve opens as the piston moves up, pushing the exhaust gases out of the cylinder.

COMPRESSION STROKE

B The piston moves up, compressing the fuel-air mixture.

INTAKE STROKE

A During the intake stroke, the piston inside the cylinder moves downward. As it does, air fills the cylinder through the intake valve, and a mist of fuel is injected into the cylinder.

POWER STROKE

C At the top of the compression stroke, a spark ignites the fuel-air mixture. The hot gases that are produced expand, pushing the piston down and turning the crankshaft.

SECTION 3 Engines and Refrigerators

INTEGRATE Career

Mechanical Engineering People who design engines and machines are mechanical engineers. Some mechanical engineers study ways to maximize the transformation of useful energy during combustion—the transformation of energy from chemical form to mechanical form.

Refrigerators

If thermal energy will only flow from something that is warm to something that is cool, how can a refrigerator be cooler inside than the air in the kitchen? A refrigerator is a heat mover. It absorbs thermal energy from the food inside the refrigerator. Then it carries the thermal energy to outside the refrigerator, where it is transferred to the surrounding air.

A refrigerator contains a material called a coolant that is pumped through pipes inside and outside the refrigerator. The coolant is the substance that carries thermal energy from the inside to the outside of the refrigerator.

Absorbing Thermal Energy Figure 14 shows how a refrigerator operates. Liquid coolant is forced up a pipe toward the freezer unit. The liquid passes through an expansion valve where it changes into a gas. When it changes into a gas, it becomes cold. The cold gas passes through pipes around the inside of the refrigerator. Because the coolant gas is so cold, it absorbs thermal energy from inside the refrigerator, and becomes warmer.

Releasing Thermal Energy However, the gas is still colder than the outside air. So, the thermal energy absorbed by the coolant cannot be transferred to the air. The coolant gas then passes through a compressor that compresses the gas. When the gas is compressed, it becomes warmer than room temperature. The gas then flows through the condenser coils, where thermal energy is transferred to the cooler air in the room. As the coolant gas cools, it changes into a liquid. The liquid is pumped through the expansion valve, changes into a gas, and the cycle is repeated.

Figure 14 A refrigerator uses a coolant to move thermal energy from inside to outside the refrigerator. The compressor supplies the energy that enables the coolant to transfer thermal energy to the room.
Diagram how the temperature of the coolant changes as it moves in a refrigerator.

Air Conditioners Most air conditioners cool in the same way that a refrigerator does. You've probably seen air-conditioning units outside of many houses. As in a refrigerator, thermal energy from inside the house is absorbed by the coolant within pipes inside the air conditioner. The coolant then is compressed by a compressor, and becomes warmer. The warmed coolant travels through pipes that are exposed to the outside air. Here the thermal energy is transferred to the outside air.

Heat Pumps Some buildings use a heat pump for heating and cooling. Like an air conditioner or refrigerator, a heat pump moves thermal energy from one place to another. In heating mode, shown in **Figure 15,** the coolant absorbs thermal energy through the outside coils. The coolant is warmed when it is compressed and transfers thermal energy to the house through the inside coils. When a heat pump is used for cooling, it removes thermal energy from the indoor air and transfers it outdoors.

Figure 15 A heat pump heats a building by absorbing thermal energy from the outside air and transferring thermal energy to the cooler air inside.

section 3 review

Summary

Heat Engines and Energy
- A heat engine is a device that converts thermal energy into mechanical energy.
- Energy cannot be created or destroyed. It only can be transformed from one form to another.
- An internal combustion engine is a heat engine that burns fuel in a combustion chamber inside the engine.

Refrigerators and Heat Pumps
- A refrigerator uses a coolant to transfer thermal energy to outside the refrigerator.
- The coolant gas absorbs thermal energy from inside the refrigerator.
- Compressing the coolant makes it warmer than the air outside the refrigerator.
- A heat pump heats by absorbing thermal energy from the air outside, and transferring it inside a building.

Self Check

1. **Diagram** the movement of coolant and the flow of heat when a heat pump is used to cool a building.
2. **Explain** why diesel engines don't use spark plugs.
3. **Identify** the source of thermal energy in an internal combustion engine.
4. **Determine** whether you could cool a kitchen by keeping the refrigerator door open.
5. **Describe** how a refrigerator uses a coolant to keep the food compartment cool.
6. **Think Critically** Explain how an air conditioner could also be used to heat a room.

Applying Skills

7. **Make a Concept Map** Make an events-chain concept map showing the sequence of steps in a four-stroke cycle.

LAB Design Your Own

Comparing Thermal Insulators

Goals
- **Predict** the temperature change of a hot drink in various types of containers over time.
- **Design** an experiment to test the hypothesis and collect data that can be graphed.
- **Interpret** the data.

Possible Materials
hot plate
large beaker
water
100-mL graduated cylinder
alcohol thermometers
various beverage containers
material to cover the containers
stopwatch
tongs
thermal gloves or mitts

Safety Precautions

WARNING: *Use caution when heating liquids. Use tongs or thermal gloves when handling hot materials. Hot and cold glass appear the same. Treat thermometers with care and keep them away from the edges of tables.*

◉ Real-World Question
Insulated beverage containers are used to reduce the flow of thermal energy. What kinds of containers do you most often drink from? Aluminum soda cans? Paper, plastic, or foam cups? Glass containers? In this investigation, compare how well several different containers reduce the transfer of thermal energy. Which types of containers are most effective at keeping a hot drink hot?

◉ Form a Hypothesis
Predict the temperature change of a hot liquid in several containers made of different materials over a time interval.

◉ Test Your Hypothesis
Make a Plan
1. **Decide** what types of containers you will test. Design an experiment to test your hypothesis. This is a group activity, so make certain that everyone gets to contribute to the discussion.

Using Scientific Methods

2. **List** the materials you will use in your experiment. Describe exactly how you will use these materials. Which liquid will you test? What will be its starting temperature? How will you cover the hot liquids in the container? What material will you use as a cover?
3. **Identify** the variables and controls in your experiment.
4. **Design** a data table in your Science Journal to record the observations you make.

Follow Your Plan

1. Have your teacher approve the steps of your experiment and your data table before you start.
2. To see the pattern of how well various containers retain thermal energy, you will need to graph your data. What kind of graph will you use? Make certain you take enough measurements during the experiment to make your graph.
3. The time intervals between measurements should be the same. Be sure to keep track of time as the experiment goes along. For how long will you measure the temperature?
4. Carry out your investigation and record your observations.

▶ Analyze Your Data

1. **Graph** your data. Use one graph to show the data collected from all your containers. Label each line on your graph.
2. **Interpret Data** How can you tell by looking at your graphs which containers best reduce the flow of thermal energy?
3. **Evaluate** Did the water temperature change as you had predicted? Use your data and graph to explain your answers.

▶ Conclude and Apply

1. **Explain** why the rate of temperature change varies among the containers. Did the size of the containers affect the rate of cooling?
2. **Conclude** which containers were the best insulators.

Compare your data and graphs with other classmates and explain any differences in your results or conclusions.

LAB **M** ◆ 175

TIME SCIENCE AND Society

SCIENCE ISSUES THAT AFFECT YOU!

The Heat Is On

You may live far from water, but still live on an island—a heat island

Think about all the things that are made of asphalt and concrete in a city. As far as the eye can see, there are buildings and parking lots, sidewalks and streets. The combined effect of these paved surfaces and towering structures can make a city sizzle in the summer. There's even a name for this effect. It's called the heat island effect.

Hot Times

You can think of a city as an island surrounded by an ocean of green trees and other vegetation. In the midst of those green trees, the air can be up to 8°C cooler than it is downtown. During the day in rural areas, the Sun's energy is absorbed by plants and soil. Some of this energy causes water to evaporate, so less energy is available to heat the surroundings. This keeps the temperature lower.

Higher temperatures aren't the only problems caused by heat islands. People crank up their air conditioners for relief, so the use of energy skyrockets. Also, the added heat speeds up the rates of chemical reactions in the atmosphere. Smog is due to chemical reactions caused by the interaction of sunlight and vehicle emissions. So hotter air means more smog. And more smog means more health problems.

Cool Cures

Several U.S. cities are working with NASA scientists to come up with a cure for the summertime blues. For instance, dark materials absorb thermal energy more efficiently than light materials. So painting buildings, especially roofs, white can reduce temperature and save on cooling bills.

Dark materials, such as asphalt, absorb more thermal energy than light materials. In extreme heat, it's even possible to fry an egg on dark pavement!

Design and Research Visit the Web Site to the right to research NASA's Urban Heat Island Project. What actions are cities taking to reduce the heat-island effect? Design a city area that would help reduce this effect.

For more information, visit
bookm.msscience.com/time

chapter 6 Study Guide

Reviewing Main Ideas

Section 1 — Temperature and Thermal Energy

1. Molecules of matter are moving constantly. Temperature is related to the average value of the kinetic energy of the molecules.
2. Thermometers measure temperature. Three common temperature scales are the Celsius, Fahrenheit, and Kelvin scales.
3. Thermal energy is the total kinetic and potential energy of the particles in matter.

Section 2 — Heat

1. Heat is thermal energy that is transferred from a warmer object to a colder object.
2. Thermal energy can be transferred by conduction, convection, and radiation.
3. A material that easily transfers thermal energy is a conductor. Thermal energy doesn't flow easily in an insulator.
4. The specific heat of a material is the thermal energy needed to change the temperature of 1 kg of the material 1°C.
5. Thermal pollution occurs when warm water is added to a body of water.

Section 3 — Engines and Refrigerators

1. A device that converts thermal energy into mechanical energy is an engine.
2. In an internal combustion engine, fuel is burned in combustion chambers inside the engine using a four-stroke cycle.
3. Refrigerators and air conditioners use a coolant to move thermal energy.

Visualizing Main Ideas

Copy and complete the following cycle map about the four-stroke cycle.

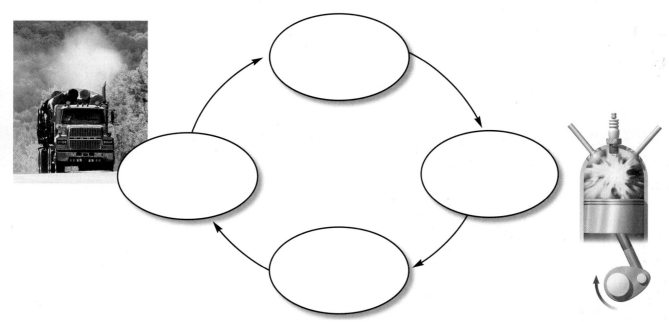

bookm.msscience.com/interactive_tutor

chapter 6 Review

Using Vocabulary

conduction p. 163
conductor p. 165
convection p. 164
heat p. 162
heat engine p. 169
internal combustion engine p. 170
radiation p. 163
specific heat p. 166
temperature p. 158
thermal energy p. 161
thermal pollution p. 167

Explain the differences in the vocabulary words given below. Then explain how the words are related. Use complete sentences in your answers.

1. internal combustion engine—heat engine
2. temperature—thermal energy
3. thermal energy—thermal pollution
4. conduction—convection
5. conduction—thermal energy
6. thermal energy—specific heat
7. conduction—radiation
8. convection—radiation
9. conductor—thermal energy

Checking Concepts

Choose the word or phrase that best answers the question.

10. What source of thermal energy does an internal combustion engine use?
    A) steam
    B) hot water
    C) burning fuel
    D) refrigerant

11. What happens to most materials when they become warmer?
    A) They contract.
    B) They float.
    C) They vaporize.
    D) They expand.

12. Which occurs if two objects at different temperatures are in contact?
    A) convection
    B) radiation
    C) condensation
    D) conduction

13. Which of the following describes the thermal energy of particles in a substance?
    A) average value of all kinetic energy
    B) total value of all kinetic energy
    C) total value of all kinetic and potential energy
    D) average value of all kinetic and potential energy

14. Thermal energy moving from the Sun to Earth is an example of which process?
    A) convection
    B) expansion
    C) radiation
    D) conduction

15. Many insulating materials contain spaces filled with air because air is what type of material?
    A) conductor
    B) coolant
    C) radiator
    D) insulator

16. A recipe calls for a cake to be baked at a temperature of 350°F. What is this temperature on the Celsius scale?
    A) 162°C
    B) 177°C
    C) 194°C
    D) 212°C

17. Which of the following is true?
    A) Warm air is less dense than cool air.
    B) Warm air is as dense as cool air.
    C) Warm air has no density.
    D) Warm air is denser than cool air.

18. Which of these is the name for thermal energy that moves from a warmer object to a cooler one?
    A) kinetic energy
    B) specific heat
    C) heat
    D) temperature

19. Which of the following is an example of thermal energy transfer by conduction?
    A) water moving in a pot of boiling water
    B) warm air rising from hot pavement
    C) the warmth you feel sitting near a fire
    D) the warmth you feel holding a cup of hot cocoa

chapter 6 Review

Thinking Critically

20. **Infer** When you heat water in a pan, the surface gets hot quickly, even though you are applying heat to the bottom of the water. Explain.

21. **Explain** why several layers of clothing often keep you warmer than a single layer.

22. **Identify** The phrase "heat rises" is sometimes used to describe the movement of thermal energy. For what type of materials is this phrase correct? Explain.

23. **Describe** When a lightbulb is turned on, the electric current in the filament causes the filament to become hot and glow. If the filament is surrounded by a gas, describe how thermal energy is transferred from the filament to the air outside the bulb.

24. **Design an Experiment** Some colors of clothing absorb radiation better than other colors. Design an experiment that will test various colors by placing them in the hot Sun for a period of time.

25. **Explain** Concrete sidewalks usually are made of slabs of concrete. Why do the concrete slabs have a space between them?

26. **Concept Map** Copy and complete the following concept map on convection in a liquid.

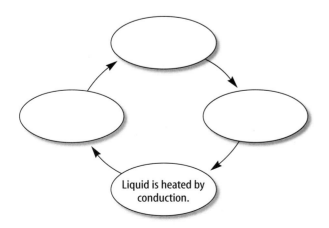

27. **Explain** A winter jacket is lined with insulating material that contains air spaces. How would the insulating properties of the jacket change if the insulating material in the jacket becomes wet? Explain.

28. **Compare** Two glasses of water are poured into a pitcher. If the temperature of the water in both glasses was the same before and after they were mixed, describe how the thermal energy of the water in the pitcher compares to the water in the glasses.

Performance Activities

29. **Poll** In the United States, the Fahrenheit temperature scale is used most often. Some people feel that Americans should switch to the Celsius scale. Take a poll of at least 20 people. Find out if they feel the switch to the Celsius scale should be made. Make a list of reasons people give for or against changing.

Applying Math

30. **Temperature Order** List the following temperatures from coldest to warmest: 80° C, 200 K, 50° F.

31. **Temperature Change** The high temperature on a summer day is 88°F and the low temperature is 61°F. What is the difference between these two temperatures in degrees Celsius?

32. **Global Temperature** The average global temperature is 286 K. Convert this temperature to degrees Celsius.

33. **Body Temperature** A doctor measures a patient's temperature at 38.4°C. Convert this temperature to degrees Fahrenheit.

chapter 6 Standardized Test Practice

Part 1 Multiple Choice

Record your answers on the answer sheet provided by your teacher or on a sheet of paper.

Use the photo below to answer questions 1 and 2.

1. The temperatures of the two glasses of water shown in the photograph above are 30°C and 0°C. Which of the following is a correct statement about the two glasses of water?
   A. The cold water has a higher average kinetic energy.
   B. The warmer water has lower thermal energy.
   C. The molecules of the cold water move faster.
   D. The molecules of the warmer water have more kinetic energy.

2. The difference in temperature of the two glasses of water is 30°C. What is their difference in temperature on the Kelvin scale?
   A. 30 K   C. 243 K
   B. 86 K   D. 303 K

3. Which of the following describes a refrigerator?
   A. heat engine   C. heat mover
   B. heat pump   D. conductor

Test-Taking Tip

Avoid rushing on test day. Prepare your clothes and test supplies the night before. Wake up early and arrive at school on time on test day.

4. Which of the following is not a step in the four-stroke cycle of internal combustion engines?
   A. compression   C. idling
   B. exhaust   D. power

Use the table below to answer question 5.

Material	Specific Heat (J/kg °C)
aluminum	897
copper	385
lead	129
nickel	444
zinc	388

5. A sample of each of the metals in the table above is formed into a 50-g cube. If 100 J of thermal energy are applied to each of the samples, which metal would change temperature by the greatest amount?
   A. aluminum   C. lead
   B. copper   D. nickel

6. An internal combustion engine converts thermal energy to which of the following forms of energy?
   A. chemical   C. radiant
   B. mechanical   D. electrical

7. Which of the following is a statement of the law of conservation of energy?
   A. Energy never can be created or destroyed.
   B. Energy can be created, but never destroyed.
   C. Energy can be destroyed, but never created.
   D. Energy can be created and destroyed when it changes form.

Standardized Test Practice

Part 2 Short Response/Grid In

Record your answers on the answer sheet provided by your teacher or on a sheet of paper.

8. If you add ice to a glass of room-temperature ice, does the water warm the ice or does the ice cool the water? Explain.

9. Strong winds that occur during a thunderstorm are the result of temperature differences between neighboring air masses. Would you expect the warmer or the cooler air mass to rise above the other?

10. A diesel engine uses a different type of fuel than the fuel used in a gasoline engine. Explain why.

11. What are the two main events that occur while the cylinder moves downward during the intake stroke of an internal combustion engine's four-stroke cycle?

Use the photo below to answer questions 12 and 13.

12. Why are cooking pots like the one in the photograph above often made of metal? Why isn't the handle made of metal?

13. When heating water in the pot, electrical energy from the cooking unit is changed to what other type of energy?

Part 3 Open Ended

Record your answers on a sheet of paper.

Use the illustration below to answer questions 14 and 15.

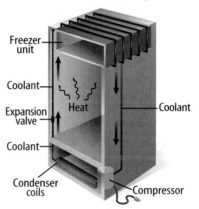

14. The illustration above shows the parts of a refrigerator and how coolant flows through the refrigerator. Explain how thermal energy is transferred to the coolant inside the refrigerator and then transferred from the coolant to the outside air.

15. What are the functions of the expansion valve, the condenser coils, and the compressor in the illustration?

16. Define convection. Explain the difference between natural and forced convection, and give an example of each.

17. Draw a sketch with arrows showing how conduction, convection, and radiation affect the movement and temperature of air near an ocean.

18. Define temperature and explain how it is related to the movement of molecules in a substance.

19. Explain what makes some materials good thermal conductors.

20. You place a cookie sheet in a hot oven. A few minutes later you hear a sound as the cookie sheet bends slightly. Explain what causes this.

Student Resources

Student Resources

CONTENTS

Science Skill Handbook 184
Scientific Methods 184
- Identify a Question 184
- Gather and Organize Information 184
- Form a Hypothesis 187
- Test the Hypothesis 188
- Collect Data 188
- Analyze the Data 191
- Draw Conclusions 192
- Communicate 192

Safety Symbols 193
Safety in the Science Laboratory 194
- General Safety Rules 194
- Prevent Accidents 194
- Laboratory Work 194
- Laboratory Cleanup 195
- Emergencies 195

Extra Try at Home Labs 196
- Measuring Momentum 196
- Friction in Traffic 196
- Submersible Egg 197
- Simple Machines 197
- The Heat is On 198
- Estimate Temperature 198

Technology Skill Handbook ...199
Computer Skills 199
- Use a Word Processing Program ...199
- Use a Database 200
- Use the Internet 200
- Use a Spreadsheet 201
- Use Graphics Software 201

Presentation Skills 202
- Develop Multimedia Presentations 202
- Computer Presentations 202

Math Skill Handbook 203
Math Review 203
- Use Fractions 203
- Use Ratios 205
- Use Decimals 205
- Use Proportions 206
- Use Percentages 206
- Solve One-Step Equations 207
- Use Statistics 208
- Use Geometry 209

Science Applications 212
- Measure in SI 212
- Dimensional Analysis 212
- Precision and Significant Digits ...214
- Scientific Notation 214
- Make and Use Graphs 215

Reference Handbooks 218
Physical Science Reference Tables 218
Periodic Table of the Elements 220

English/Spanish Glossary 222

Index 228

Credits 233

Scientific Methods

Scientists use an orderly approach called the scientific method to solve problems. This includes organizing and recording data so others can understand them. Scientists use many variations in this method when they solve problems.

Identify a Question

The first step in a scientific investigation or experiment is to identify a question to be answered or a problem to be solved. For example, you might ask which gasoline is the most efficient.

Gather and Organize Information

After you have identified your question, begin gathering and organizing information. There are many ways to gather information, such as researching in a library, interviewing those knowledgeable about the subject, testing and working in the laboratory and field. Fieldwork is investigations and observations done outside of a laboratory.

Researching Information Before moving in a new direction, it is important to gather the information that already is known about the subject. Start by asking yourself questions to determine exactly what you need to know. Then you will look for the information in various reference sources, like the student is doing in **Figure 1.** Some sources may include textbooks, encyclopedias, government documents, professional journals, science magazines, and the Internet. Always list the sources of your information.

Figure 1 The Internet can be a valuable research tool.

Evaluate Sources of Information Not all sources of information are reliable. You should evaluate all of your sources of information, and use only those you know to be dependable. For example, if you are researching ways to make homes more energy efficient, a site written by the U.S. Department of Energy would be more reliable than a site written by a company that is trying to sell a new type of weatherproofing material. Also, remember that research always is changing. Consult the most current resources available to you. For example, a 1985 resource about saving energy would not reflect the most recent findings.

Sometimes scientists use data that they did not collect themselves, or conclusions drawn by other researchers. This data must be evaluated carefully. Ask questions about how the data were obtained, if the investigation was carried out properly, and if it has been duplicated exactly with the same results. Would you reach the same conclusion from the data? Only when you have confidence in the data can you believe it is true and feel comfortable using it.

Science Skill Handbook

Interpret Scientific Illustrations As you research a topic in science, you will see drawings, diagrams, and photographs to help you understand what you read. Some illustrations are included to help you understand an idea that you can't see easily by yourself, like the tiny particles in an atom in **Figure 2.** A drawing helps many people to remember details more easily and provides examples that clarify difficult concepts or give additional information about the topic you are studying. Most illustrations have labels or a caption to identify or to provide more information.

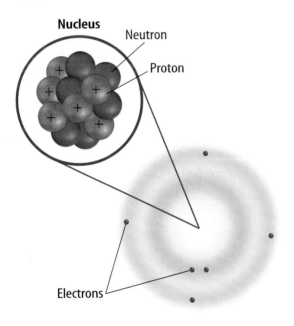

Figure 2 This drawing shows an atom of carbon with its six protons, six neutrons, and six electrons.

Concept Maps One way to organize data is to draw a diagram that shows relationships among ideas (or concepts). A concept map can help make the meanings of ideas and terms more clear, and help you understand and remember what you are studying. Concept maps are useful for breaking large concepts down into smaller parts, making learning easier.

Network Tree A type of concept map that not only shows a relationship, but how the concepts are related is a network tree, shown in **Figure 3.** In a network tree, the words are written in the ovals, while the description of the type of relationship is written across the connecting lines.

When constructing a network tree, write down the topic and all major topics on separate pieces of paper or notecards. Then arrange them in order from general to specific. Branch the related concepts from the major concept and describe the relationship on the connecting line. Continue to more specific concepts until finished.

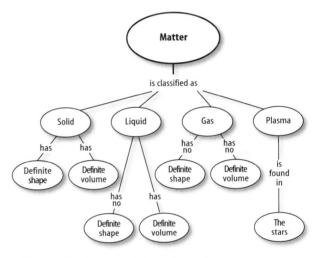

Figure 3 A network tree shows how concepts or objects are related.

Events Chain Another type of concept map is an events chain. Sometimes called a flow chart, it models the order or sequence of items. An events chain can be used to describe a sequence of events, the steps in a procedure, or the stages of a process.

When making an events chain, first find the one event that starts the chain. This event is called the initiating event. Then, find the next event and continue until the outcome is reached, as shown in **Figure 4.**

SCIENCE SKILL HANDBOOK M ◆ 185

Science Skill Handbook

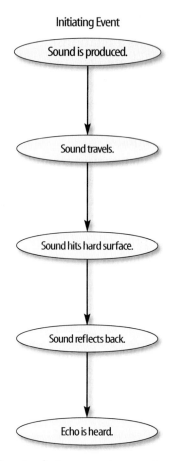

Figure 4 Events-chain concept maps show the order of steps in a process or event. This concept map shows how a sound makes an echo.

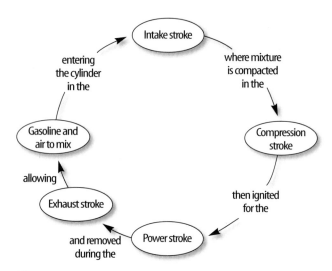

Figure 5 A cycle map shows events that occur in a cycle.

Cycle Map A specific type of events chain is a cycle map. It is used when the series of events do not produce a final outcome, but instead relate back to the beginning event, such as in **Figure 5**. Therefore, the cycle repeats itself.

To make a cycle map, first decide what event is the beginning event. This is also called the initiating event. Then list the next events in the order that they occur, with the last event relating back to the initiating event. Words can be written between the events that describe what happens from one event to the next. The number of events in a cycle map can vary, but usually contain three or more events.

Spider Map A type of concept map that you can use for brainstorming is the spider map. When you have a central idea, you might find that you have a jumble of ideas that relate to it but are not necessarily clearly related to each other. The spider map on sound in **Figure 6** shows that if you write these ideas outside the main concept, then you can begin to separate and group unrelated terms so they become more useful.

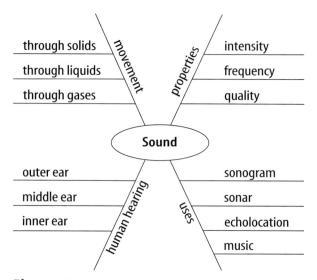

Figure 6 A spider map allows you to list ideas that relate to a central topic but not necessarily to one another.

186 ◆ M STUDENT RESOURCES

Science Skill Handbook

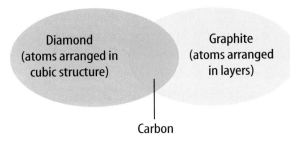

Figure 7 This Venn diagram compares and contrasts two substances made from carbon.

Venn Diagram To illustrate how two subjects compare and contrast you can use a Venn diagram. You can see the characteristics that the subjects have in common and those that they do not, shown in **Figure 7**.

To create a Venn diagram, draw two overlapping ovals that that are big enough to write in. List the characteristics unique to one subject in one oval, and the characteristics of the other subject in the other oval. The characteristics in common are listed in the overlapping section.

Make and Use Tables One way to organize information so it is easier to understand is to use a table. Tables can contain numbers, words, or both.

To make a table, list the items to be compared in the first column and the characteristics to be compared in the first row. The title should clearly indicate the content of the table, and the column or row heads should be clear. Notice that in **Table 1** the units are included.

Table 1 Recyclables Collected During Week			
Day of Week	Paper (kg)	Aluminum (kg)	Glass (kg)
Monday	5.0	4.0	12.0
Wednesday	4.0	1.0	10.0
Friday	2.5	2.0	10.0

Make a Model One way to help you better understand the parts of a structure, the way a process works, or to show things too large or small for viewing is to make a model. For example, an atomic model made of a plastic-ball nucleus and pipe-cleaner electron shells can help you visualize how the parts of an atom relate to each other. Other types of models can by devised on a computer or represented by equations.

Form a Hypothesis

A possible explanation based on previous knowledge and observations is called a hypothesis. After researching gasoline types and recalling previous experiences in your family's car you form a hypothesis—our car runs more efficiently because we use premium gasoline. To be valid, a hypothesis has to be something you can test by using an investigation.

Predict When you apply a hypothesis to a specific situation, you predict something about that situation. A prediction makes a statement in advance, based on prior observation, experience, or scientific reasoning. People use predictions to make everyday decisions. Scientists test predictions by performing investigations. Based on previous observations and experiences, you might form a prediction that cars are more efficient with premium gasoline. The prediction can be tested in an investigation.

Design an Experiment A scientist needs to make many decisions before beginning an investigation. Some of these include: how to carry out the investigation, what steps to follow, how to record the data, and how the investigation will answer the question. It also is important to address any safety concerns.

Science Skill Handbook

Test the Hypothesis

Now that you have formed your hypothesis, you need to test it. Using an investigation, you will make observations and collect data, or information. This data might either support or not support your hypothesis. Scientists collect and organize data as numbers and descriptions.

Follow a Procedure In order to know what materials to use, as well as how and in what order to use them, you must follow a procedure. **Figure 8** shows a procedure you might follow to test your hypothesis.

> **Procedure**
> 1. Use regular gasoline for two weeks.
> 2. Record the number of kilometers between fill-ups and the amount of gasoline used.
> 3. Switch to premium gasoline for two weeks.
> 4. Record the number of kilometers between fill-ups and the amount of gasoline used.

Figure 8 A procedure tells you what to do step by step.

Identify and Manipulate Variables and Controls In any experiment, it is important to keep everything the same except for the item you are testing. The one factor you change is called the independent variable. The change that results is the dependent variable. Make sure you have only one independent variable, to assure yourself of the cause of the changes you observe in the dependent variable. For example, in your gasoline experiment the type of fuel is the independent variable. The dependent variable is the efficiency.

Many experiments also have a control—an individual instance or experimental subject for which the independent variable is not changed. You can then compare the test results to the control results. To design a control you can have two cars of the same type. The control car uses regular gasoline for four weeks. After you are done with the test, you can compare the experimental results to the control results.

Collect Data

Whether you are carrying out an investigation or a short observational experiment, you will collect data, as shown in **Figure 9**. Scientists collect data as numbers and descriptions and organize it in specific ways.

Observe Scientists observe items and events, then record what they see. When they use only words to describe an observation, it is called qualitative data. Scientists' observations also can describe how much there is of something. These observations use numbers, as well as words, in the description and are called quantitative data. For example, if a sample of the element gold is described as being "shiny and very dense" the data are qualitative. Quantitative data on this sample of gold might include "a mass of 30 g and a density of 19.3 g/cm^3."

Figure 9 Collecting data is one way to gather information directly.

Science Skill Handbook

Figure 10 Record data neatly and clearly so it is easy to understand.

When you make observations you should examine the entire object or situation first, and then look carefully for details. It is important to record observations accurately and completely. Always record your notes immediately as you make them, so you do not miss details or make a mistake when recording results from memory. Never put unidentified observations on scraps of paper. Instead they should be recorded in a notebook, like the one in **Figure 10.** Write your data neatly so you can easily read it later. At each point in the experiment, record your observations and label them. That way, you will not have to determine what the figures mean when you look at your notes later. Set up any tables that you will need to use ahead of time, so you can record any observations right away. Remember to avoid bias when collecting data by not including personal thoughts when you record observations. Record only what you observe.

Estimate Scientific work also involves estimating. To estimate is to make a judgment about the size or the number of something without measuring or counting. This is important when the number or size of an object or population is too large or too difficult to accurately count or measure.

Sample Scientists may use a sample or a portion of the total number as a type of estimation. To sample is to take a small, representative portion of the objects or organisms of a population for research. By making careful observations or manipulating variables within that portion of the group, information is discovered and conclusions are drawn that might apply to the whole population. A poorly chosen sample can be unrepresentative of the whole. If you were trying to determine the rainfall in an area, it would not be best to take a rainfall sample from under a tree.

Measure You use measurements everyday. Scientists also take measurements when collecting data. When taking measurements, it is important to know how to use measuring tools properly. Accuracy also is important.

Length To measure length, the distance between two points, scientists use meters. Smaller measurements might be measured in centimeters or millimeters.

Length is measured using a metric ruler or meter stick. When using a metric ruler, line up the 0-cm mark with the end of the object being measured and read the number of the unit where the object ends. Look at the metric ruler shown in **Figure 11.** The centimeter lines are the long, numbered lines, and the shorter lines are millimeter lines. In this instance, the length would be 4.50 cm.

Figure 11 This metric ruler has centimeter and millimeter divisions.

SCIENCE SKILL HANDBOOK M ◆ 189

Science Skill Handbook

Mass The SI unit for mass is the kilogram (kg). Scientists can measure mass using units formed by adding metric prefixes to the unit gram (g), such as milligram (mg). To measure mass, you might use a triple-beam balance similar to the one shown in **Figure 12.** The balance has a pan on one side and a set of beams on the other side. Each beam has a rider that slides on the beam.

When using a triple-beam balance, place an object on the pan. Slide the largest rider along its beam until the pointer drops below zero. Then move it back one notch. Repeat the process for each rider proceeding from the larger to smaller until the pointer swings an equal distance above and below the zero point. Sum the masses on each beam to find the mass of the object. Move all riders back to zero when finished.

Instead of putting materials directly on the balance, scientists often take a tare of a container. A tare is the mass of a container into which objects or substances are placed for measuring their masses. To mass objects or substances, find the mass of a clean container. Remove the container from the pan, and place the object or substances in the container. Find the mass of the container with the materials in it. Subtract the mass of the empty container from the mass of the filled container to find the mass of the materials you are using.

Figure 12 A triple-beam balance is used to determine the mass of an object.

Figure 13 Graduated cylinders measure liquid volume.

Liquid Volume To measure liquids, the unit used is the liter. When a smaller unit is needed, scientists might use a milliliter. Because a milliliter takes up the volume of a cube measuring 1 cm on each side it also can be called a cubic centimeter ($cm^3 = cm \times cm \times cm$).

You can use beakers and graduated cylinders to measure liquid volume. A graduated cylinder, shown in **Figure 13,** is marked from bottom to top in milliliters. In lab, you might use a 10-mL graduated cylinder or a 100-mL graduated cylinder. When measuring liquids, notice that the liquid has a curved surface. Look at the surface at eye level, and measure the bottom of the curve. This is called the meniscus. The graduated cylinder in **Figure 13** contains 79.0 mL, or 79.0 cm^3, of a liquid.

Temperature Scientists often measure temperature using the Celsius scale. Pure water has a freezing point of 0°C and boiling point of 100°C. The unit of measurement is degrees Celsius. Two other scales often used are the Fahrenheit and Kelvin scales.

Science Skill Handbook

Figure 14 A thermometer measures the temperature of an object.

Scientists use a thermometer to measure temperature. Most thermometers in a laboratory are glass tubes with a bulb at the bottom end containing a liquid such as colored alcohol. The liquid rises or falls with a change in temperature. To read a glass thermometer like the thermometer in **Figure 14,** rotate it slowly until a red line appears. Read the temperature where the red line ends.

Form Operational Definitions An operational definition defines an object by how it functions, works, or behaves. For example, when you are playing hide and seek and a tree is home base, you have created an operational definition for a tree.

Objects can have more than one operational definition. For example, a ruler can be defined as a tool that measures the length of an object (how it is used). It can also be a tool with a series of marks used as a standard when measuring (how it works).

Analyze the Data

To determine the meaning of your observations and investigation results, you will need to look for patterns in the data. Then you must think critically to determine what the data mean. Scientists use several approaches when they analyze the data they have collected and recorded. Each approach is useful for identifying specific patterns.

Interpret Data The word *interpret* means "to explain the meaning of something." When analyzing data from an experiment, try to find out what the data show. Identify the control group and the test group to see whether or not changes in the independent variable have had an effect. Look for differences in the dependent variable between the control and test groups.

Classify Sorting objects or events into groups based on common features is called classifying. When classifying, first observe the objects or events to be classified. Then select one feature that is shared by some members in the group, but not by all. Place those members that share that feature in a subgroup. You can classify members into smaller and smaller subgroups based on characteristics. Remember that when you classify, you are grouping objects or events for a purpose. Keep your purpose in mind as you select the features to form groups and subgroups.

Compare and Contrast Observations can be analyzed by noting the similarities and differences between two more objects or events that you observe. When you look at objects or events to see how they are similar, you are comparing them. Contrasting is looking for differences in objects or events.

SCIENCE SKILL HANDBOOK M ♦ 191

Science Skill Handbook

Recognize Cause and Effect A cause is a reason for an action or condition. The effect is that action or condition. When two events happen together, it is not necessarily true that one event caused the other. Scientists must design a controlled investigation to recognize the exact cause and effect.

Draw Conclusions

When scientists have analyzed the data they collected, they proceed to draw conclusions about the data. These conclusions are sometimes stated in words similar to the hypothesis that you formed earlier. They may confirm a hypothesis, or lead you to a new hypothesis.

Infer Scientists often make inferences based on their observations. An inference is an attempt to explain observations or to indicate a cause. An inference is not a fact, but a logical conclusion that needs further investigation. For example, you may infer that a fire has caused smoke. Until you investigate, however, you do not know for sure.

Apply When you draw a conclusion, you must apply those conclusions to determine whether the data supports the hypothesis. If your data do not support your hypothesis, it does not mean that the hypothesis is wrong. It means only that the result of the investigation did not support the hypothesis. Maybe the experiment needs to be redesigned, or some of the initial observations on which the hypothesis was based were incomplete or biased. Perhaps more observation or research is needed to refine your hypothesis. A successful investigation does not always come out the way you originally predicted.

Avoid Bias Sometimes a scientific investigation involves making judgments. When you make a judgment, you form an opinion. It is important to be honest and not to allow any expectations of results to bias your judgments. This is important throughout the entire investigation, from researching to collecting data to drawing conclusions.

Communicate

The communication of ideas is an important part of the work of scientists. A discovery that is not reported will not advance the scientific community's understanding or knowledge. Communication among scientists also is important as a way of improving their investigations.

Scientists communicate in many ways, from writing articles in journals and magazines that explain their investigations and experiments, to announcing important discoveries on television and radio. Scientists also share ideas with colleagues on the Internet or present them as lectures, like the student is doing in **Figure 15**.

Figure 15 A student communicates to his peers about his investigation.

Science Skill Handbook

SAFETY SYMBOLS

	HAZARD	EXAMPLES	PRECAUTION	REMEDY
DISPOSAL	Special disposal procedures need to be followed.	certain chemicals, living organisms	Do not dispose of these materials in the sink or trash can.	Dispose of wastes as directed by your teacher.
BIOLOGICAL	Organisms or other biological materials that might be harmful to humans	bacteria, fungi, blood, unpreserved tissues, plant materials	Avoid skin contact with these materials. Wear mask or gloves.	Notify your teacher if you suspect contact with material. Wash hands thoroughly.
EXTREME TEMPERATURE	Objects that can burn skin by being too cold or too hot	boiling liquids, hot plates, dry ice, liquid nitrogen	Use proper protection when handling.	Go to your teacher for first aid.
SHARP OBJECT	Use of tools or glassware that can easily puncture or slice skin	razor blades, pins, scalpels, pointed tools, dissecting probes, broken glass	Practice common-sense behavior and follow guidelines for use of the tool.	Go to your teacher for first aid.
FUME	Possible danger to respiratory tract from fumes	ammonia, acetone, nail polish remover, heated sulfur, moth balls	Make sure there is good ventilation. Never smell fumes directly. Wear a mask.	Leave foul area and notify your teacher immediately.
ELECTRICAL	Possible danger from electrical shock or burn	improper grounding, liquid spills, short circuits, exposed wires	Double-check setup with teacher. Check condition of wires and apparatus.	Do not attempt to fix electrical problems. Notify your teacher immediately.
IRRITANT	Substances that can irritate the skin or mucous membranes of the respiratory tract	pollen, moth balls, steel wool, fiberglass, potassium permanganate	Wear dust mask and gloves. Practice extra care when handling these materials.	Go to your teacher for first aid.
CHEMICAL	Chemicals can react with and destroy tissue and other materials	bleaches such as hydrogen peroxide; acids such as sulfuric acid, hydrochloric acid; bases such as ammonia, sodium hydroxide	Wear goggles, gloves, and an apron.	Immediately flush the affected area with water and notify your teacher.
TOXIC	Substance may be poisonous if touched, inhaled, or swallowed.	mercury, many metal compounds, iodine, poinsettia plant parts	Follow your teacher's instructions.	Always wash hands thoroughly after use. Go to your teacher for first aid.
FLAMMABLE	Flammable chemicals may be ignited by open flame, spark, or exposed heat.	alcohol, kerosene, potassium permanganate	Avoid open flames and heat when using flammable chemicals.	Notify your teacher immediately. Use fire safety equipment if applicable.
OPEN FLAME	Open flame in use, may cause fire.	hair, clothing, paper, synthetic materials	Tie back hair and loose clothing. Follow teacher's instruction on lighting and extinguishing flames.	Notify your teacher immediately. Use fire safety equipment if applicable.

 Eye Safety Proper eye protection should be worn at all times by anyone performing or observing science activities.

 Clothing Protection This symbol appears when substances could stain or burn clothing.

 Animal Safety This symbol appears when safety of animals and students must be ensured.

 Handwashing After the lab, wash hands with soap and water before removing goggles.

Safety in the Science Laboratory

The science laboratory is a safe place to work if you follow standard safety procedures. Being responsible for your own safety helps to make the entire laboratory a safer place for everyone. When performing any lab, read and apply the caution statements and safety symbol listed at the beginning of the lab.

General Safety Rules

1. Obtain your teacher's permission to begin all investigations and use laboratory equipment.

2. Study the procedure. Ask your teacher any questions. Be sure you understand safety symbols shown on the page.

3. Notify your teacher about allergies or other health conditions which can affect your participation in a lab.

4. Learn and follow use and safety procedures for your equipment. If unsure, ask your teacher.

5. Never eat, drink, chew gum, apply cosmetics, or do any personal grooming in the lab. Never use lab glassware as food or drink containers. Keep your hands away from your face and mouth.

6. Know the location and proper use of the safety shower, eye wash, fire blanket, and fire alarm.

Prevent Accidents

1. Use the safety equipment provided to you. Goggles and a safety apron should be worn during investigations.

2. Do NOT use hair spray, mousse, or other flammable hair products. Tie back long hair and tie down loose clothing.

3. Do NOT wear sandals or other open-toed shoes in the lab.

4. Remove jewelry on hands and wrists. Loose jewelry, such as chains and long necklaces, should be removed to prevent them from getting caught in equipment.

5. Do not taste any substances or draw any material into a tube with your mouth.

6. Proper behavior is expected in the lab. Practical jokes and fooling around can lead to accidents and injury.

7. Keep your work area uncluttered.

Laboratory Work

1. Collect and carry all equipment and materials to your work area before beginning a lab.

2. Remain in your own work area unless given permission by your teacher to leave it.

Science Skill Handbook

3. Dispose of chemicals and other materials as directed by your teacher. Place broken glass and solid substances in the proper containers. Never discard materials in the sink.

4. Clean your work area.

5. Wash your hands with soap and water thoroughly BEFORE removing your goggles.

Emergencies

1. Report any fire, electrical shock, glassware breakage, spill, or injury, no matter how small, to your teacher immediately. Follow his or her instructions.

2. If your clothing should catch fire, STOP, DROP, and ROLL. If possible, smother it with the fire blanket or get under a safety shower. NEVER RUN.

3. If a fire should occur, turn off all gas and leave the room according to established procedures.

4. In most instances, your teacher will clean up spills. Do NOT attempt to clean up spills unless you are given permission and instructions to do so.

5. If chemicals come into contact with your eyes or skin, notify your teacher immediately. Use the eyewash or flush your skin or eyes with large quantities of water.

6. The fire extinguisher and first-aid kit should only be used by your teacher unless it is an extreme emergency and you have been given permission.

7. If someone is injured or becomes ill, only a professional medical provider or someone certified in first aid should perform first-aid procedures.

3. Always slant test tubes away from yourself and others when heating them, adding substances to them, or rinsing them.

4. If instructed to smell a substance in a container, hold the container a short distance away and fan vapors towards your nose.

5. Do NOT substitute other chemicals/substances for those in the materials list unless instructed to do so by your teacher.

6. Do NOT take any materials or chemicals outside of the laboratory.

7. Stay out of storage areas unless instructed to be there and supervised by your teacher.

Laboratory Cleanup

1. Turn off all burners, water, and gas, and disconnect all electrical devices.

2. Clean all pieces of equipment and return all materials to their proper places.

Extra Try at Home Labs

EXTRA Try at Home Labs

From Your Kitchen, Junk Drawer, or Yard

1 Measuring Momentum

Real-World Question
How much momentum do rolling balls have?

Possible Materials
- meterstick
- orange cones or tape
- scale
- stopwatch
- bucket
- bowling ball
- plastic baseball
- golf ball
- tennis ball
- calculator

Procedure
1. Use a balance to measure the masses of the tennis ball, golf ball, and plastic baseball. Convert their masses from grams to kilograms.
2. Find the weight of the bowling ball in pounds. The weight should be written on the ball. Divide the ball's weight by 2.2 to calculate its mass in kilograms.
3. Go outside and measure a 10-m distance on a blacktop or concrete surface. Mark the distance with orange cones or tape.
4. Have a partner roll each ball the 10-m distance. Measure the time it takes each ball to roll 10 m.
5. Use the formula: $\text{velocity} = \frac{\text{distance}}{\text{time}}$ to calculate each ball's velocity.

Conclude and Apply
1. Calculate the momentum of each ball.
2. Infer why the momentums of the balls differed so greatly.

2 Friction in Traffic

Real-World Question
How do the various kinds of friction affect the operation of vehicles?

Possible Materials
- erasers taken from the ends of pencils (4)
- needles (2)
- small match box
- toy car

Procedure
1. Build a match box car with the materials listed, or use a toy car.
2. Invent ways to demonstrate the effects of static friction, sliding friction, and rolling friction on the car. Think of hills, ice or rain conditions, graveled roads and paved roads, etc.
3. Make drawings of how friction is acting on the car, or how the car uses friction to work.

Conclude and Apply
1. In what ways are static, sliding, and rolling friction helpful to drivers?
2. In what ways are static, sliding, and rolling friction unfavorable to car safety and operation?
3. Explain what your experiment taught you about driving in icy conditions.

Adult supervision required for all labs.

Extra Try at Home Labs

3 Submersible Egg

Real-World Question
How can you make an egg float and sink again?

Possible Materials
- egg
- 10 mL (2 teaspoons) of salt
- measuring spoons
- water
- glass (250–300 mL)
- spoon
- marking pen

Procedure
1. Put 150 mL of water in the glass.
2. Gently use a marking pen to write an X on one side of the egg.
3. Put the egg in the glass. Record your observations.
4. Use the spoon to remove the egg. Add 1 mL (1/4 teaspoon) of salt and swirl or stir to dissolve. Put the egg back in. Record your observations.
5. Repeat step 4 until you have used all the salt. Remember to make observations at each step.

Conclude and Apply
1. How could you resink the egg? Try your idea. Did it work?
2. The egg always floats with the same point or side down. Why do you think this is?
3. How does your answer to question 2 relate to real-world applications?

4 Simple Machines

Real-World Question
What types of simple machines are found in a toolbox?

Possible Materials
- box of tools

Procedure
1. Obtain a box of tools and lay all the tools and other hardware from the box on a table.
2. Carefully examine all the tools and hardware, and separate all the items that are a type of inclined plane.
3. Carefully examine all the tools and hardware, and separate all the items that are a type of lever.
4. Identify and separate all the items that are a wheel and axle.
5. Identify any pulleys in the toolbox.
6. Identify any tools that are a combination of two or more simple machines.

Conclude and Apply
1. List all the tools you found that were a type of inclined plane, lever, wheel and axle, or pulley.
2. List all the tools that were a combination of two or more simple machines.
3. Infer how a hammer could be used as both a first class lever and a third class lever.

Adult supervision required for all labs.

Extra Try at Home Labs

5 The Heat is On

Real-World Question
How can different types of energy be transformed into thermal energy?

Possible Materials
- lamp
- incandescent light bulb
- black construction paper or cloth

Procedure
1. Feel the temperature of a black sheet of paper. Lay the paper in direct sunlight, wait 10 min, and observe how it feels.
2. Rub the palms of your hands together quickly for 10 s and observe how they feel.
3. Switch on a lamp that has a bare light bulb. *Without touching the lightbulb,* cup your hand 2 cm above the bulb for 30 s and observe what you feel.

Conclude and Apply
1. Infer the type of energy transformation that happened on the paper.
2. Infer the type of energy transformation that happened between the palms of your hands.
3. Infer the type of energy transformation that happened to the lightbulb.

6 Estimate Temperature

Real-World Question
How can we learn to estimate temperatures?

Possible Materials
- thermometer
- bowl
- water
- ice

Procedure
1. If you have a dual-scale weather thermometer, you can learn twice as much by trying to do your estimation in degrees Fahrenheit and Celsius each time.
2. Fill a bowl with ice water. Submerge your fingers in the water and estimate the water temperature.
3. Place the thermometer in the bowl and observe the temperature.
4. Place a bowl of warm water in direct sunlight for 20 min. Submerge your fingers in the water and estimate the water temperature.
5. Place the thermometer in the bowl and observe the temperature.
6. Place the thermometer outside in a location where you can see it each day.
7. Each day for a month, step outside and estimate the temperature. Check the accuracy of your estimates with the thermometer. Record the weather conditions as well.

Conclude and Apply
1. Describe how well you can estimate air temperatures after estimating the temperature each day for a month. Did the cloudiness of the day affect your estimation skills?
2. Infer why understanding the Celsius scale might be helpful to you in the future.

Adult supervision required for all labs.

Technology Skill Handbook

Computer Skills

People who study science rely on computers, like the one in **Figure 16,** to record and store data and to analyze results from investigations. Whether you work in a laboratory or just need to write a lab report with tables, good computer skills are a necessity.

Using the computer comes with responsibility. Issues of ownership, security, and privacy can arise. Remember, if you did not author the information you are using, you must provide a source for your information. Also, anything on a computer can be accessed by others. Do not put anything on the computer that you would not want everyone to know. To add more security to your work, use a password.

Use a Word Processing Program

A computer program that allows you to type your information, change it as many times as you need to, and then print it out is called a word processing program. Word processing programs also can be used to make tables.

Learn the Skill To start your word processing program, a blank document, sometimes called "Document 1," appears on the screen. To begin, start typing. To create a new document, click the *New* button on the standard tool bar. These tips will help you format the document.

- The program will automatically move to the next line; press *Enter* if you wish to start a new paragraph.
- Symbols, called non-printing characters, can be hidden by clicking the *Show/Hide* button on your toolbar.
- To insert text, move the cursor to the point where you want the insertion to go, click on the mouse once, and type the text.
- To move several lines of text, select the text and click the *Cut* button on your toolbar. Then position your cursor in the location that you want to move the cut text and click *Paste*. If you move to the wrong place, click *Undo*.
- The spell check feature does not catch words that are misspelled to look like other words, like "cold" instead of "gold." Always reread your document to catch all spelling mistakes.
- To learn about other word processing methods, read the user's manual or click on the *Help* button.
- You can integrate databases, graphics, and spreadsheets into documents by copying from another program and pasting it into your document, or by using desktop publishing (DTP). DTP software allows you to put text and graphics together to finish your document with a professional look. This software varies in how it is used and its capabilities.

Figure 16 A computer will make reports neater and more professional looking.

TECHNOLOGY SKILL HANDBOOK M ◆ 199

Technology Skill Handbook

Use a Database

A collection of facts stored in a computer and sorted into different fields is called a database. A database can be reorganized in any way that suits your needs.

Learn the Skill A computer program that allows you to create your own database is a database management system (DBMS). It allows you to add, delete, or change information. Take time to get to know the features of your database software.

- Determine what facts you would like to include and research to collect your information.
- Determine how you want to organize the information.
- Follow the instructions for your particular DBMS to set up fields. Then enter each item of data in the appropriate field.
- Follow the instructions to sort the information in order of importance.
- Evaluate the information in your database, and add, delete, or change as necessary.

Use the Internet

The Internet is a global network of computers where information is stored and shared. To use the Internet, like the students in **Figure 17,** you need a modem to connect your computer to a phone line and an Internet Service Provider account.

Learn the Skill To access internet sites and information, use a "Web browser," which lets you view and explore pages on the World Wide Web. Each page is its own site, and each site has its own address, called a URL. Once you have found a Web browser, follow these steps for a search (this also is how you search a database).

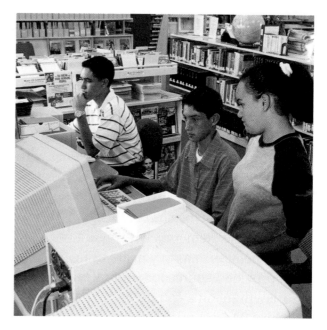

Figure 17 The Internet allows you to search a global network for a variety of information.

- Be as specific as possible. If you know you want to research "gold," don't type in "elements." Keep narrowing your search until you find what you want.
- Web sites that end in *.com* are commercial Web sites; *.org, .edu,* and *.gov* are nonprofit, educational, or government Web sites.
- Electronic encyclopedias, almanacs, indexes, and catalogs will help locate and select relevant information.
- Develop a "home page" with relative ease. When developing a Web site, NEVER post pictures or disclose personal information such as location, names, or phone numbers. Your school or community usually can host your Web site. A basic understanding of HTML (hypertext mark-up language), the language of Web sites, is necessary. Software that creates HTML code is called authoring software, and can be downloaded free from many Web sites. This software allows text and pictures to be arranged as the software is writing the HTML code.

Use a Spreadsheet

A spreadsheet, shown in **Figure 18,** can perform mathematical functions with any data arranged in columns and rows. By entering a simple equation into a cell, the program can perform operations in specific cells, rows, or columns.

Learn the Skill Each column (vertical) is assigned a letter, and each row (horizontal) is assigned a number. Each point where a row and column intersect is called a cell, and is labeled according to where it is located—Column A, Row 1 (A1).

- Decide how to organize the data, and enter it in the correct row or column.
- Spreadsheets can use standard formulas or formulas can be customized to calculate cells.
- To make a change, click on a cell to make it activate, and enter the edited data or formula.
- Spreadsheets also can display your results in graphs. Choose the style of graph that best represents the data.

Figure 18 A spreadsheet allows you to perform mathematical operations on your data.

Use Graphics Software

Adding pictures, called graphics, to your documents is one way to make your documents more meaningful and exciting. This software adds, edits, and even constructs graphics. There is a variety of graphics software programs. The tools used for drawing can be a mouse, keyboard, or other specialized devices. Some graphics programs are simple. Others are complicated, called computer-aided design (CAD) software.

Learn the Skill It is important to have an understanding of the graphics software being used before starting. The better the software is understood, the better the results. The graphics can be placed in a word-processing document.

- Clip art can be found on a variety of internet sites, and on CDs. These images can be copied and pasted into your document.
- When beginning, try editing existing drawings, then work up to creating drawings.
- The images are made of tiny rectangles of color called pixels. Each pixel can be altered.
- Digital photography is another way to add images. The photographs in the memory of a digital camera can be downloaded into a computer, then edited and added to the document.
- Graphics software also can allow animation. The software allows drawings to have the appearance of movement by connecting basic drawings automatically. This is called in-betweening, or tweening.
- Remember to save often.

Technology Skill Handbook

Presentation Skills

Develop Multimedia Presentations

Most presentations are more dynamic if they include diagrams, photographs, videos, or sound recordings, like the one shown in **Figure 19.** A multimedia presentation involves using stereos, overhead projectors, televisions, computers, and more.

Learn the Skill Decide the main points of your presentation, and what types of media would best illustrate those points.

- Make sure you know how to use the equipment you are working with.
- Practice the presentation using the equipment several times.
- Enlist the help of a classmate to push play or turn lights out for you. Be sure to practice your presentation with him or her.
- If possible, set up all of the equipment ahead of time, and make sure everything is working properly.

Figure 19 These students are engaging the audience using a variety of tools.

Computer Presentations

There are many different interactive computer programs that you can use to enhance your presentation. Most computers have a compact disc (CD) drive that can play both CDs and digital video discs (DVDs). Also, there is hardware to connect a regular CD, DVD, or VCR. These tools will enhance your presentation.

Another method of using the computer to aid in your presentation is to develop a slide show using a computer program. This can allow movement of visuals at the presenter's pace, and can allow for visuals to build on one another.

Learn the Skill In order to create multimedia presentations on a computer, you need to have certain tools. These may include traditional graphic tools and drawing programs, animation programs, and authoring systems that tie everything together. Your computer will tell you which tools it supports. The most important step is to learn about the tools that you will be using.

- Often, color and strong images will convey a point better than words alone. Use the best methods available to convey your point.
- As with other presentations, practice many times.
- Practice your presentation with the tools you and any assistants will be using.
- Maintain eye contact with the audience. The purpose of using the computer is not to prompt the presenter, but to help the audience understand the points of the presentation.

Math Review

Use Fractions

A fraction compares a part to a whole. In the fraction $\frac{2}{3}$, the 2 represents the part and is the numerator. The 3 represents the whole and is the denominator.

Reduce Fractions To reduce a fraction, you must find the largest factor that is common to both the numerator and the denominator, the greatest common factor (GCF). Divide both numbers by the GCF. The fraction has then been reduced, or it is in its simplest form.

Example Twelve of the 20 chemicals in the science lab are in powder form. What fraction of the chemicals used in the lab are in powder form?

Step 1 Write the fraction.
$$\frac{part}{whole} = \frac{12}{20}$$

Step 2 To find the GCF of the numerator and denominator, list all of the factors of each number.
Factors of 12: 1, 2, 3, 4, 6, 12 (the numbers that divide evenly into 12)
Factors of 20: 1, 2, 4, 5, 10, 20 (the numbers that divide evenly into 20)

Step 3 List the common factors.
1, 2, 4.

Step 4 Choose the greatest factor in the list.
The GCF of 12 and 20 is 4.

Step 5 Divide the numerator and denominator by the GCF.
$$\frac{12 \div 4}{20 \div 4} = \frac{3}{5}$$

In the lab, $\frac{3}{5}$ of the chemicals are in powder form.

Practice Problem At an amusement park, 66 of 90 rides have a height restriction. What fraction of the rides, in its simplest form, has a height restriction?

Add and Subtract Fractions To add or subtract fractions with the same denominator, add or subtract the numerators and write the sum or difference over the denominator. After finding the sum or difference, find the simplest form for your fraction.

Example 1 In the forest outside your house, $\frac{1}{8}$ of the animals are rabbits, $\frac{3}{8}$ are squirrels, and the remainder are birds and insects. How many are mammals?

Step 1 Add the numerators.
$$\frac{1}{8} + \frac{3}{8} = \frac{(1+3)}{8} = \frac{4}{8}$$

Step 2 Find the GCF.
$$\frac{4}{8} \text{ (GCF, 4)}$$

Step 3 Divide the numerator and denominator by the GCF.
$$\frac{4}{4} = 1, \frac{8}{4} = 2$$

$\frac{1}{2}$ of the animals are mammals.

Example 2 If $\frac{7}{16}$ of the Earth is covered by freshwater, and $\frac{1}{16}$ of that is in glaciers, how much freshwater is not frozen?

Step 1 Subtract the numerators.
$$\frac{7}{16} - \frac{1}{16} = \frac{(7-1)}{16} = \frac{6}{16}$$

Step 2 Find the GCF.
$$\frac{6}{16} \text{ (GCF, 2)}$$

Step 3 Divide the numerator and denominator by the GCF.
$$\frac{6}{2} = 3, \frac{16}{2} = 8$$

$\frac{3}{8}$ of the freshwater is not frozen.

Practice Problem A bicycle rider is going 15 km/h for $\frac{4}{9}$ of his ride, 10 km/h for $\frac{2}{9}$ of his ride, and 8 km/h for the remainder of the ride. How much of his ride is he going over 8 km/h?

Math Skill Handbook

Unlike Denominators To add or subtract fractions with unlike denominators, first find the least common denominator (LCD). This is the smallest number that is a common multiple of both denominators. Rename each fraction with the LCD, and then add or subtract. Find the simplest form if necessary.

Example 1 A chemist makes a paste that is $\frac{1}{2}$ table salt (NaCl), $\frac{1}{3}$ sugar ($C_6H_{12}O_6$), and the rest water (H_2O). How much of the paste is a solid?

Step 1 Find the LCD of the fractions.
$\frac{1}{2} + \frac{1}{3}$ (LCD, 6)

Step 2 Rename each numerator and each denominator with the LCD.
$1 \times 3 = 3, \ 2 \times 3 = 6$
$1 \times 2 = 2, \ 3 \times 2 = 6$

Step 3 Add the numerators.
$\frac{3}{6} + \frac{2}{6} = \frac{(3+2)}{6} = \frac{5}{6}$

$\frac{5}{6}$ of the paste is a solid.

Example 2 The average precipitation in Grand Junction, CO, is $\frac{7}{10}$ inch in November, and $\frac{3}{5}$ inch in December. What is the total average precipitation?

Step 1 Find the LCD of the fractions.
$\frac{7}{10} + \frac{3}{5}$ (LCD, 10)

Step 2 Rename each numerator and each denominator with the LCD.
$7 \times 1 = 7, \ 10 \times 1 = 10$
$3 \times 2 = 6, \ 5 \times 2 = 10$

Step 3 Add the numerators.
$\frac{7}{10} + \frac{6}{10} = \frac{(7+6)}{10} = \frac{13}{10}$

$\frac{13}{10}$ inches total precipitation, or $1\frac{3}{10}$ inches.

Practice Problem On an electric bill, about $\frac{1}{8}$ of the energy is from solar energy and about $\frac{1}{10}$ is from wind power. How much of the total bill is from solar energy and wind power combined?

Example 3 In your body, $\frac{7}{10}$ of your muscle contractions are involuntary (cardiac and smooth muscle tissue). Smooth muscle makes $\frac{3}{15}$ of your muscle contractions. How many of your muscle contractions are made by cardiac muscle?

Step 1 Find the LCD of the fractions.
$\frac{7}{10} - \frac{3}{15}$ (LCD, 30)

Step 2 Rename each numerator and each denominator with the LCD.
$7 \times 3 = 21, \ 10 \times 3 = 30$
$3 \times 2 = 6, \ 15 \times 2 = 30$

Step 3 Subtract the numerators.
$\frac{21}{30} - \frac{6}{30} = \frac{(21-6)}{30} = \frac{15}{30}$

Step 4 Find the GCF.
$\frac{15}{30}$ (GCF, 15)
$\frac{1}{2}$

$\frac{1}{2}$ of all muscle contractions are cardiac muscle.

Example 4 Tony wants to make cookies that call for $\frac{3}{4}$ of a cup of flour, but he only has $\frac{1}{3}$ of a cup. How much more flour does he need?

Step 1 Find the LCD of the fractions.
$\frac{3}{4} - \frac{1}{3}$ (LCD, 12)

Step 2 Rename each numerator and each denominator with the LCD.
$3 \times 3 = 9, \ 4 \times 3 = 12$
$1 \times 4 = 4, \ 3 \times 4 = 12$

Step 3 Subtract the numerators.
$\frac{9}{12} - \frac{4}{12} = \frac{(9-4)}{12} = \frac{5}{12}$

$\frac{5}{12}$ of a cup of flour.

Practice Problem Using the information provided to you in Example 3 above, determine how many muscle contractions are voluntary (skeletal muscle).

Math Skill Handbook

Multiply Fractions To multiply with fractions, multiply the numerators and multiply the denominators. Find the simplest form if necessary.

Example Multiply $\frac{3}{5}$ by $\frac{1}{3}$.

Step 1 Multiply the numerators and denominators.
$$\frac{3}{5} \times \frac{1}{3} = \frac{(3 \times 1)}{(5 \times 3)} = \frac{3}{15}$$

Step 2 Find the GCF.
$\frac{3}{15}$ (GCF, 3)

Step 3 Divide the numerator and denominator by the GCF.
$\frac{3}{3} = 1, \frac{15}{3} = 5$
$\frac{1}{5}$

$\frac{3}{5}$ multiplied by $\frac{1}{3}$ is $\frac{1}{5}$.

Practice Problem Multiply $\frac{3}{14}$ by $\frac{5}{16}$.

Find a Reciprocal Two numbers whose product is 1 are called multiplicative inverses, or reciprocals.

Example Find the reciprocal of $\frac{3}{8}$.

Step 1 Inverse the fraction by putting the denominator on top and the numerator on the bottom.
$\frac{8}{3}$

The reciprocal of $\frac{3}{8}$ is $\frac{8}{3}$.

Practice Problem Find the reciprocal of $\frac{4}{9}$.

Divide Fractions To divide one fraction by another fraction, multiply the dividend by the reciprocal of the divisor. Find the simplest form if necessary.

Example 1 Divide $\frac{1}{9}$ by $\frac{1}{3}$.

Step 1 Find the reciprocal of the divisor.
The reciprocal of $\frac{1}{3}$ is $\frac{3}{1}$.

Step 2 Multiply the dividend by the reciprocal of the divisor.
$$\frac{\frac{1}{9}}{\frac{1}{3}} = \frac{1}{9} \times \frac{3}{1} = \frac{(1 \times 3)}{(9 \times 1)} = \frac{3}{9}$$

Step 3 Find the GCF.
$\frac{3}{9}$ (GCF, 3)

Step 4 Divide the numerator and denominator by the GCF.
$\frac{3}{3} = 1, \frac{9}{3} = 3$
$\frac{1}{3}$

$\frac{1}{9}$ divided by $\frac{1}{3}$ is $\frac{1}{3}$.

Example 2 Divide $\frac{3}{5}$ by $\frac{1}{4}$.

Step 1 Find the reciprocal of the divisor.
The reciprocal of $\frac{1}{4}$ is $\frac{4}{1}$.

Step 2 Multiply the dividend by the reciprocal of the divisor.
$$\frac{\frac{3}{5}}{\frac{1}{4}} = \frac{3}{5} \times \frac{4}{1} = \frac{(3 \times 4)}{(5 \times 1)} = \frac{12}{5}$$

$\frac{3}{5}$ divided by $\frac{1}{4}$ is $\frac{12}{5}$ or $2\frac{2}{5}$.

Practice Problem Divide $\frac{3}{11}$ by $\frac{7}{10}$.

Math Skill Handbook

Use Ratios

When you compare two numbers by division, you are using a ratio. Ratios can be written 3 to 5, 3:5, or $\frac{3}{5}$. Ratios, like fractions, also can be written in simplest form.

Ratios can represent probabilities, also called odds. This is a ratio that compares the number of ways a certain outcome occurs to the number of outcomes. For example, if you flip a coin 100 times, what are the odds that it will come up heads? There are two possible outcomes, heads or tails, so the odds of coming up heads are 50:100. Another way to say this is that 50 out of 100 times the coin will come up heads. In its simplest form, the ratio is 1:2.

Example 1 A chemical solution contains 40 g of salt and 64 g of baking soda. What is the ratio of salt to baking soda as a fraction in simplest form?

Step 1 Write the ratio as a fraction.
$$\frac{\text{salt}}{\text{baking soda}} = \frac{40}{64}$$

Step 2 Express the fraction in simplest form.
The GCF of 40 and 64 is 8.
$$\frac{40}{64} = \frac{40 \div 8}{64 \div 8} = \frac{5}{8}$$

The ratio of salt to baking soda in the sample is 5:8.

Example 2 Sean rolls a 6-sided die 6 times. What are the odds that the side with a 3 will show?

Step 1 Write the ratio as a fraction.
$$\frac{\text{number of sides with a 3}}{\text{number of sides}} = \frac{1}{6}$$

Step 2 Multiply by the number of attempts.
$$\frac{1}{6} \times 6 \text{ attempts} = \frac{6}{6} \text{ attempts} = 1 \text{ attempt}$$

1 attempt out of 6 will show a 3.

Practice Problem Two metal rods measure 100 cm and 144 cm in length. What is the ratio of their lengths in simplest form?

Use Decimals

A fraction with a denominator that is a power of ten can be written as a decimal. For example, 0.27 means $\frac{27}{100}$. The decimal point separates the ones place from the tenths place.

Any fraction can be written as a decimal using division. For example, the fraction $\frac{5}{8}$ can be written as a decimal by dividing 5 by 8. Written as a decimal, it is 0.625.

Add or Subtract Decimals When adding and subtracting decimals, line up the decimal points before carrying out the operation.

Example 1 Find the sum of 47.68 and 7.80.

Step 1 Line up the decimal places when you write the numbers.
```
  47.68
+  7.80
```

Step 2 Add the decimals.
```
  47.68
+  7.80
  -----
  55.48
```

The sum of 47.68 and 7.80 is 55.48.

Example 2 Find the difference of 42.17 and 15.85.

Step 1 Line up the decimal places when you write the number.
```
  42.17
- 15.85
```

Step 2 Subtract the decimals.
```
  42.17
- 15.85
  -----
  26.32
```

The difference of 42.17 and 15.85 is 26.32.

Practice Problem Find the sum of 1.245 and 3.842.

Math Skill Handbook

Multiply Decimals To multiply decimals, multiply the numbers like any other number, ignoring the decimal point. Count the decimal places in each factor. The product will have the same number of decimal places as the sum of the decimal places in the factors.

Example Multiply 2.4 by 5.9.

Step 1 Multiply the factors like two whole numbers.
$24 \times 59 = 1416$

Step 2 Find the sum of the number of decimal places in the factors. Each factor has one decimal place, for a sum of two decimal places.

Step 3 The product will have two decimal places.
14.16

The product of 2.4 and 5.9 is 14.16.

Practice Problem Multiply 4.6 by 2.2.

Divide Decimals When dividing decimals, change the divisor to a whole number. To do this, multiply both the divisor and the dividend by the same power of ten. Then place the decimal point in the quotient directly above the decimal point in the dividend. Then divide as you do with whole numbers.

Example Divide 8.84 by 3.4.

Step 1 Multiply both factors by 10.
$3.4 \times 10 = 34$, $8.84 \times 10 = 88.4$

Step 2 Divide 88.4 by 34.

```
      2.6
34)88.4
   -68
    204
   -204
      0
```

8.84 divided by 3.4 is 2.6.

Practice Problem Divide 75.6 by 3.6.

Use Proportions

An equation that shows that two ratios are equivalent is a proportion. The ratios $\frac{2}{4}$ and $\frac{5}{10}$ are equivalent, so they can be written as $\frac{2}{4} = \frac{5}{10}$. This equation is a proportion.

When two ratios form a proportion, the cross products are equal. To find the cross products in the proportion $\frac{2}{4} = \frac{5}{10}$, multiply the 2 and the 10, and the 4 and the 5. Therefore $2 \times 10 = 4 \times 5$, or $20 = 20$.

Because you know that both proportions are equal, you can use cross products to find a missing term in a proportion. This is known as solving the proportion.

Example The heights of a tree and a pole are proportional to the lengths of their shadows. The tree casts a shadow of 24 m when a 6-m pole casts a shadow of 4 m. What is the height of the tree?

Step 1 Write a proportion.
$$\frac{\text{height of tree}}{\text{height of pole}} = \frac{\text{length of tree's shadow}}{\text{length of pole's shadow}}$$

Step 2 Substitute the known values into the proportion. Let h represent the unknown value, the height of the tree.
$$\frac{h}{6} = \frac{24}{4}$$

Step 3 Find the cross products.
$h \times 4 = 6 \times 24$

Step 4 Simplify the equation.
$4h = 144$

Step 5 Divide each side by 4.
$$\frac{4h}{4} = \frac{144}{4}$$
$h = 36$

The height of the tree is 36 m.

Practice Problem The ratios of the weights of two objects on the Moon and on Earth are in proportion. A rock weighing 3 N on the Moon weighs 18 N on Earth. How much would a rock that weighs 5 N on the Moon weigh on Earth?

Math Skill Handbook

Use Percentages

The word *percent* means "out of one hundred." It is a ratio that compares a number to 100. Suppose you read that 77 percent of the Earth's surface is covered by water. That is the same as reading that the fraction of the Earth's surface covered by water is $\frac{77}{100}$. To express a fraction as a percent, first find the equivalent decimal for the fraction. Then, multiply the decimal by 100 and add the percent symbol.

Example Express $\frac{13}{20}$ as a percent.

Step 1 Find the equivalent decimal for the fraction.

```
      0.65
20)13.00
   12 0
    1 00
    1 00
       0
```

Step 2 Rewrite the fraction $\frac{13}{20}$ as 0.65.

Step 3 Multiply 0.65 by 100 and add the % sign.
$0.65 \times 100 = 65 = 65\%$

So, $\frac{13}{20} = 65\%$.

This also can be solved as a proportion.

Example Express $\frac{13}{20}$ as a percent.

Step 1 Write a proportion.
$\frac{13}{20} = \frac{x}{100}$

Step 2 Find the cross products.
$1300 = 20x$

Step 3 Divide each side by 20.
$\frac{1300}{20} = \frac{20x}{20}$
$65\% = x$

Practice Problem In one year, 73 of 365 days were rainy in one city. What percent of the days in that city were rainy?

Solve One-Step Equations

A statement that two things are equal is an equation. For example, $A = B$ is an equation that states that A is equal to B.

An equation is solved when a variable is replaced with a value that makes both sides of the equation equal. To make both sides equal the inverse operation is used. Addition and subtraction are inverses, and multiplication and division are inverses.

Example 1 Solve the equation $x - 10 = 35$.

Step 1 Find the solution by adding 10 to each side of the equation.
$x - 10 = 35$
$x - 10 + 10 = 35 + 10$
$x = 45$

Step 2 Check the solution.
$x - 10 = 35$
$45 - 10 = 35$
$35 = 35$

Both sides of the equation are equal, so $x = 45$.

Example 2 In the formula $a = bc$, find the value of c if $a = 20$ and $b = 2$.

Step 1 Rearrange the formula so the unknown value is by itself on one side of the equation by dividing both sides by b.
$a = bc$
$\frac{a}{b} = \frac{bc}{b}$
$\frac{a}{b} = c$

Step 2 Replace the variables a and b with the values that are given.
$\frac{a}{b} = c$
$\frac{20}{2} = c$
$10 = c$

Step 3 Check the solution.
$a = bc$
$20 = 2 \times 10$
$20 = 20$

Both sides of the equation are equal, so $c = 10$ is the solution when $a = 20$ and $b = 2$.

Practice Problem In the formula $h = gd$, find the value of d if $g = 12.3$ and $h = 17.4$.

Math Skill Handbook

Use Statistics

The branch of mathematics that deals with collecting, analyzing, and presenting data is statistics. In statistics, there are three common ways to summarize data with a single number—the mean, the median, and the mode.

The **mean** of a set of data is the arithmetic average. It is found by adding the numbers in the data set and dividing by the number of items in the set.

The **median** is the middle number in a set of data when the data are arranged in numerical order. If there were an even number of data points, the median would be the mean of the two middle numbers.

The **mode** of a set of data is the number or item that appears most often.

Another number that often is used to describe a set of data is the range. The **range** is the difference between the largest number and the smallest number in a set of data.

A **frequency table** shows how many times each piece of data occurs, usually in a survey. **Table 2** below shows the results of a student survey on favorite color.

Table 2 Student Color Choice		
Color	Tally	Frequency
red	IIII	4
blue	HHT	5
black	II	2
green	III	3
purple	HHT II	7
yellow	HHT I	6

Based on the frequency table data, which color is the favorite?

Example The speeds (in m/s) for a race car during five different time trials are 39, 37, 44, 36, and 44.

To find the mean:

Step 1 Find the sum of the numbers.
39 + 37 + 44 + 36 + 44 = 200

Step 2 Divide the sum by the number of items, which is 5.
200 ÷ 5 = 40

The mean is 40 m/s.

To find the median:

Step 1 Arrange the measures from least to greatest.
36, 37, 39, 44, 44

Step 2 Determine the middle measure.
36, 37, <u>39</u>, 44, 44

The median is 39 m/s.

To find the mode:

Step 1 Group the numbers that are the same together.
44, 44, 36, 37, 39

Step 2 Determine the number that occurs most in the set.
<u>44, 44</u>, 36, 37, 39

The mode is 44 m/s.

To find the range:

Step 1 Arrange the measures from largest to smallest.
44, 44, 39, 37, 36

Step 2 Determine the largest and smallest measures in the set.
<u>44</u>, 44, 39, 37, <u>36</u>

Step 3 Find the difference between the largest and smallest measures.
44 − 36 = 8

The range is 8 m/s.

Practice Problem Find the mean, median, mode, and range for the data set 8, 4, 12, 8, 11, 14, 16.

Math Skill Handbook

Use Geometry

The branch of mathematics that deals with the measurement, properties, and relationships of points, lines, angles, surfaces, and solids is called geometry.

Perimeter The **perimeter** (P) is the distance around a geometric figure. To find the perimeter of a rectangle, add the length and width and multiply that sum by two, or $2(l + w)$. To find perimeters of irregular figures, add the length of the sides.

Example 1 Find the perimeter of a rectangle that is 3 m long and 5 m wide.

Step 1 You know that the perimeter is 2 times the sum of the width and length.
$P = 2(3\text{ m} + 5\text{ m})$

Step 2 Find the sum of the width and length.
$P = 2(8\text{ m})$

Step 3 Multiply by 2.
$P = 16\text{ m}$

The perimeter is 16 m.

Example 2 Find the perimeter of a shape with sides measuring 2 cm, 5 cm, 6 cm, 3 cm.

Step 1 You know that the perimeter is the sum of all the sides.
$P = 2 + 5 + 6 + 3$

Step 2 Find the sum of the sides.
$P = 2 + 5 + 6 + 3$
$P = 16$

The perimeter is 16 cm.

Practice Problem Find the perimeter of a rectangle with a length of 18 m and a width of 7 m.

Practice Problem Find the perimeter of a triangle measuring 1.6 cm by 2.4 cm by 2.4 cm.

Area of a Rectangle The **area** (A) is the number of square units needed to cover a surface. To find the area of a rectangle, multiply the length times the width, or $l \times w$. When finding area, the units also are multiplied. Area is given in square units.

Example Find the area of a rectangle with a length of 1 cm and a width of 10 cm.

Step 1 You know that the area is the length multiplied by the width.
$A = (1\text{ cm} \times 10\text{ cm})$

Step 2 Multiply the length by the width. Also multiply the units.
$A = 10\text{ cm}^2$

The area is 10 cm^2.

Practice Problem Find the area of a square whose sides measure 4 m.

Area of a Triangle To find the area of a triangle, use the formula:

$$A = \frac{1}{2}(\text{base} \times \text{height})$$

The base of a triangle can be any of its sides. The height is the perpendicular distance from a base to the opposite endpoint, or vertex.

Example Find the area of a triangle with a base of 18 m and a height of 7 m.

Step 1 You know that the area is $\frac{1}{2}$ the base times the height.
$A = \frac{1}{2}(18\text{ m} \times 7\text{ m})$

Step 2 Multiply $\frac{1}{2}$ by the product of 18×7. Multiply the units.
$A = \frac{1}{2}(126\text{ m}^2)$
$A = 63\text{ m}^2$

The area is 63 m^2.

Practice Problem Find the area of a triangle with a base of 27 cm and a height of 17 cm.

Math Skill Handbook

Circumference of a Circle The **diameter** (*d*) of a circle is the distance across the circle through its center, and the **radius** (*r*) is the distance from the center to any point on the circle. The radius is half of the diameter. The distance around the circle is called the **circumference** (C). The formula for finding the circumference is:

$C = 2\pi r$ *or* $C = \pi d$

The circumference divided by the diameter is always equal to 3.1415926... This nonterminating and nonrepeating number is represented by the Greek letter π (pi). An approximation often used for π is 3.14.

Example 1 Find the circumference of a circle with a radius of 3 m.

Step 1 You know the formula for the circumference is 2 times the radius times π.
$C = 2\pi(3)$

Step 2 Multiply 2 times the radius.
$C = 6\pi$

Step 3 Multiply by π.
$C = 19$ m

The circumference is 19 m.

Example 2 Find the circumference of a circle with a diameter of 24.0 cm.

Step 1 You know the formula for the circumference is the diameter times π.
$C = \pi(24.0)$

Step 2 Multiply the diameter by π.
$C = 75.4$ cm

The circumference is 75.4 cm.

Practice Problem Find the circumference of a circle with a radius of 19 cm.

Area of a Circle The formula for the area of a circle is:
$A = \pi r^2$

Example 1 Find the area of a circle with a radius of 4.0 cm.

Step 1 $A = \pi(4.0)^2$

Step 2 Find the square of the radius.
$A = 16\pi$

Step 3 Multiply the square of the radius by π.
$A = 50$ cm^2

The area of the circle is 50 cm^2.

Example 2 Find the area of a circle with a radius of 225 m.

Step 1 $A = \pi(225)^2$

Step 2 Find the square of the radius.
$A = 50625\pi$

Step 3 Multiply the square of the radius by π.
$A = 158962.5$

The area of the circle is 158,962 m^2.

Example 3 Find the area of a circle whose diameter is 20.0 mm.

Step 1 You know the formula for the area of a circle is the square of the radius times π, and that the radius is half of the diameter.
$A = \pi\left(\dfrac{20.0}{2}\right)^2$

Step 2 Find the radius.
$A = \pi(10.0)^2$

Step 3 Find the square of the radius.
$A = 100\pi$

Step 4 Multiply the square of the radius by π.
$A = 314$ mm^2

The area is 314 mm^2.

Practice Problem Find the area of a circle with a radius of 16 m.

Math Skill Handbook

Volume The measure of space occupied by a solid is the **volume** (V). To find the volume of a rectangular solid multiply the length times width times height, or $V = l \times w \times h$. It is measured in cubic units, such as cubic centimeters (cm^3).

Example Find the volume of a rectangular solid with a length of 2.0 m, a width of 4.0 m, and a height of 3.0 m.

Step 1 You know the formula for volume is the length times the width times the height.
$V = 2.0 \text{ m} \times 4.0 \text{ m} \times 3.0 \text{ m}$

Step 2 Multiply the length times the width times the height.
$V = 24 \text{ m}^3$

The volume is 24 m^3.

Practice Problem Find the volume of a rectangular solid that is 8 m long, 4 m wide, and 4 m high.

To find the volume of other solids, multiply the area of the base times the height.

Example 1 Find the volume of a solid that has a triangular base with a length of 8.0 m and a height of 7.0 m. The height of the entire solid is 15.0 m.

Step 1 You know that the base is a triangle, and the area of a triangle is $\frac{1}{2}$ the base times the height, and the volume is the area of the base times the height.
$V = \left[\frac{1}{2}(b \times h)\right] \times 15$

Step 2 Find the area of the base.
$V = \left[\frac{1}{2}(8 \times 7)\right] \times 15$
$V = \left(\frac{1}{2} \times 56\right) \times 15$

Step 3 Multiply the area of the base by the height of the solid.
$V = 28 \times 15$
$V = 420 \text{ m}^3$

The volume is 420 m^3.

Example 2 Find the volume of a cylinder that has a base with a radius of 12.0 cm, and a height of 21.0 cm.

Step 1 You know that the base is a circle, and the area of a circle is the square of the radius times π, and the volume is the area of the base times the height.
$V = (\pi r^2) \times 21$
$V = (\pi 12^2) \times 21$

Step 2 Find the area of the base.
$V = 144\pi \times 21$
$V = 452 \times 21$

Step 3 Multiply the area of the base by the height of the solid.
$V = 9490 \text{ cm}^3$

The volume is 9490 cm^3.

Example 3 Find the volume of a cylinder that has a diameter of 15 mm and a height of 4.8 mm.

Step 1 You know that the base is a circle with an area equal to the square of the radius times π. The radius is one-half the diameter. The volume is the area of the base times the height.
$V = (\pi r^2) \times 4.8$
$V = \left[\pi\left(\frac{1}{2} \times 15\right)^2\right] \times 4.8$
$V = (\pi 7.5^2) \times 4.8$

Step 2 Find the area of the base.
$V = 56.25\pi \times 4.8$
$V = 176.63 \times 4.8$

Step 3 Multiply the area of the base by the height of the solid.
$V = 847.8$

The volume is 847.8 mm^3.

Practice Problem Find the volume of a cylinder with a diameter of 7 cm in the base and a height of 16 cm.

Math Skill Handbook

Science Applications

Measure in SI

The metric system of measurement was developed in 1795. A modern form of the metric system, called the International System (SI), was adopted in 1960 and provides the standard measurements that all scientists around the world can understand.

The SI system is convenient because unit sizes vary by powers of 10. Prefixes are used to name units. Look at **Table 3** for some common SI prefixes and their meanings.

Table 3 Common SI Prefixes			
Prefix	Symbol	Meaning	
kilo-	k	1,000	thousand
hecto-	h	100	hundred
deka-	da	10	ten
deci-	d	0.1	tenth
centi-	c	0.01	hundredth
milli-	m	0.001	thousandth

Example How many grams equal one kilogram?

Step 1 Find the prefix *kilo* in **Table 3**.

Step 2 Using **Table 3**, determine the meaning of *kilo*. According to the table, it means 1,000. When the prefix *kilo* is added to a unit, it means that there are 1,000 of the units in a "*kilo*unit."

Step 3 Apply the prefix to the units in the question. The units in the question are grams. There are 1,000 grams in a kilogram.

Practice Problem Is a milligram larger or smaller than a gram? How many of the smaller units equal one larger unit? What fraction of the larger unit does one smaller unit represent?

Dimensional Analysis

Convert SI Units In science, quantities such as length, mass, and time sometimes are measured using different units. A process called dimensional analysis can be used to change one unit of measure to another. This process involves multiplying your starting quantity and units by one or more conversion factors. A conversion factor is a ratio equal to one and can be made from any two equal quantities with different units. If 1,000 mL equal 1 L then two ratios can be made.

$$\frac{1,000 \text{ mL}}{1 \text{ L}} = \frac{1 \text{ L}}{1,000 \text{ mL}} = 1$$

One can covert between units in the SI system by using the equivalents in **Table 3** to make conversion factors.

Example 1 How many cm are in 4 m?

Step 1 Write conversion factors for the units given. From **Table 3,** you know that 100 cm = 1 m. The conversion factors are

$$\frac{100 \text{ cm}}{1 \text{ m}} \text{ and } \frac{1 \text{ m}}{100 \text{ cm}}$$

Step 2 Decide which conversion factor to use. Select the factor that has the units you are converting from (m) in the denominator and the units you are converting to (cm) in the numerator.

$$\frac{100 \text{ cm}}{1 \text{ m}}$$

Step 3 Multiply the starting quantity and units by the conversion factor. Cancel the starting units with the units in the denominator. There are 400 cm in 4 m.

$$4 \text{ m} \times \frac{100 \text{ cm}}{1 \text{ m}} = 400 \text{ cm}$$

Practice Problem How many milligrams are in one kilogram? (Hint: You will need to use two conversion factors from **Table 3**.)

Math Skill Handbook

Table 4 Unit System Equivalents

Type of Measurement	Equivalent
Length	1 in = 2.54 cm
	1 yd = 0.91 m
	1 mi = 1.61 km
Mass and Weight*	1 oz = 28.35 g
	1 lb = 0.45 kg
	1 ton (short) = 0.91 tonnes (metric tons)
	1 lb = 4.45 N
Volume	1 in^3 = 16.39 cm^3
	1 qt = 0.95 L
	1 gal = 3.78 L
Area	1 in^2 = 6.45 cm^2
	1 yd^2 = 0.83 m^2
	1 mi^2 = 2.59 km^2
	1 acre = 0.40 hectares
Temperature	°C = $\frac{(°F - 32)}{1.8}$
	K = °C + 273

*Weight is measured in standard Earth gravity.

Convert Between Unit Systems Table 4 gives a list of equivalents that can be used to convert between English and SI units.

Example If a meterstick has a length of 100 cm, how long is the meterstick in inches?

Step 1 Write the conversion factors for the units given. From **Table 4,** 1 in = 2.54 cm.

$$\frac{1 \text{ in}}{2.54 \text{ cm}} \text{ and } \frac{2.54 \text{ cm}}{1 \text{ in}}$$

Step 2 Determine which conversion factor to use. You are converting from cm to in. Use the conversion factor with cm on the bottom.

$$\frac{1 \text{ in}}{2.54 \text{ cm}}$$

Step 3 Multiply the starting quantity and units by the conversion factor. Cancel the starting units with the units in the denominator. Round your answer based on the number of significant figures in the conversion factor.

$$100 \text{ cm} \times \frac{1 \text{ in}}{2.54 \text{ cm}} = 39.37 \text{ in}$$

The meterstick is 39.4 in long.

Practice Problem A book has a mass of 5 lbs. What is the mass of the book in kg?

Practice Problem Use the equivalent for in and cm (1 in = 2.54 cm) to show how 1 in^3 = 16.39 cm^3.

Math Skill Handbook

Precision and Significant Digits

When you make a measurement, the value you record depends on the precision of the measuring instrument. This precision is represented by the number of significant digits recorded in the measurement. When counting the number of significant digits, all digits are counted except zeros at the end of a number with no decimal point such as 2,050, and zeros at the beginning of a decimal such as 0.03020. When adding or subtracting numbers with different precision, round the answer to the smallest number of decimal places of any number in the sum or difference. When multiplying or dividing, the answer is rounded to the smallest number of significant digits of any number being multiplied or divided.

Example The lengths 5.28 and 5.2 are measured in meters. Find the sum of these lengths and record your answer using the correct number of significant digits.

Step 1 Find the sum.

 5.28 m   2 digits after the decimal
+ 5.2 m   1 digit after the decimal
─────────
10.48 m

Step 2 Round to one digit after the decimal because the least number of digits after the decimal of the numbers being added is 1.

The sum is 10.5 m.

Practice Problem How many significant digits are in the measurement 7,071,301 m? How many significant digits are in the measurement 0.003010 g?

Practice Problem Multiply 5.28 and 5.2 using the rule for multiplying and dividing. Record the answer using the correct number of significant digits.

Scientific Notation

Many times numbers used in science are very small or very large. Because these numbers are difficult to work with scientists use scientific notation. To write numbers in scientific notation, move the decimal point until only one non-zero digit remains on the left. Then count the number of places you moved the decimal point and use that number as a power of ten. For example, the average distance from the Sun to Mars is 227,800,000,000 m. In scientific notation, this distance is 2.278×10^{11} m. Because you moved the decimal point to the left, the number is a positive power of ten.

The mass of an electron is about 0.000 000 000 000 000 000 000 000 000 000 911 kg. Expressed in scientific notation, this mass is 9.11×10^{-31} kg. Because the decimal point was moved to the right, the number is a negative power of ten.

Example Earth is 149,600,000 km from the Sun. Express this in scientific notation.

Step 1 Move the decimal point until one non-zero digit remains on the left.
1.496 000 00

Step 2 Count the number of decimal places you have moved. In this case, eight.

Step 3 Show that number as a power of ten, 10^8.

The Earth is 1.496×10^8 km from the Sun.

Practice Problem How many significant digits are in 149,600,000 km? How many significant digits are in 1.496×10^8 km?

Practice Problem Parts used in a high performance car must be measured to 7×10^{-6} m. Express this number as a decimal.

Practice Problem A CD is spinning at 539 revolutions per minute. Express this number in scientific notation.

Math Skill Handbook

Make and Use Graphs

Data in tables can be displayed in a graph—a visual representation of data. Common graph types include line graphs, bar graphs, and circle graphs.

Line Graph A line graph shows a relationship between two variables that change continuously. The independent variable is changed and is plotted on the *x*-axis. The dependent variable is observed, and is plotted on the *y*-axis.

Example Draw a line graph of the data below from a cyclist in a long-distance race.

Table 5 Bicycle Race Data	
Time (h)	Distance (km)
0	0
1	8
2	16
3	24
4	32
5	40

Step 1 Determine the *x*-axis and *y*-axis variables. Time varies independently of distance and is plotted on the *x*-axis. Distance is dependent on time and is plotted on the *y*-axis.

Step 2 Determine the scale of each axis. The *x*-axis data ranges from 0 to 5. The *y*-axis data ranges from 0 to 40.

Step 3 Using graph paper, draw and label the axes. Include units in the labels.

Step 4 Draw a point at the intersection of the time value on the *x*-axis and corresponding distance value on the *y*-axis. Connect the points and label the graph with a title, as shown in **Figure 20**.

Figure 20 This line graph shows the relationship between distance and time during a bicycle ride.

Practice Problem A puppy's shoulder height is measured during the first year of her life. The following measurements were collected: (3 mo, 52 cm), (6 mo, 72 cm), (9 mo, 83 cm), (12 mo, 86 cm). Graph this data.

Find a Slope The slope of a straight line is the ratio of the vertical change, rise, to the horizontal change, run.

$$\text{Slope} = \frac{\text{vertical change (rise)}}{\text{horizontal change (run)}} = \frac{\text{change in } y}{\text{change in } x}$$

Example Find the slope of the graph in **Figure 20**.

Step 1 You know that the slope is the change in *y* divided by the change in *x*.
$$\text{Slope} = \frac{\text{change in } y}{\text{change in } x}$$

Step 2 Determine the data points you will be using. For a straight line, choose the two sets of points that are the farthest apart.
$$\text{Slope} = \frac{(40-0) \text{ km}}{(5-0) \text{ hr}}$$

Step 3 Find the change in *y* and *x*.
$$\text{Slope} = \frac{40 \text{ km}}{5 \text{ h}}$$

Step 4 Divide the change in *y* by the change in *x*.
$$\text{Slope} = \frac{8 \text{ km}}{\text{h}}$$

The slope of the graph is 8 km/h.

Math Skill Handbook

Bar Graph To compare data that does not change continuously you might choose a bar graph. A bar graph uses bars to show the relationships between variables. The *x*-axis variable is divided into parts. The parts can be numbers such as years, or a category such as a type of animal. The *y*-axis is a number and increases continuously along the axis.

Example A recycling center collects 4.0 kg of aluminum on Monday, 1.0 kg on Wednesday, and 2.0 kg on Friday. Create a bar graph of this data.

Step 1 Select the *x*-axis and *y*-axis variables. The measured numbers (the masses of aluminum) should be placed on the *y*-axis. The variable divided into parts (collection days) is placed on the *x*-axis.

Step 2 Create a graph grid like you would for a line graph. Include labels and units.

Step 3 For each measured number, draw a vertical bar above the *x*-axis value up to the *y*-axis value. For the first data point, draw a vertical bar above Monday up to 4.0 kg.

Practice Problem Draw a bar graph of the gases in air: 78% nitrogen, 21% oxygen, 1% other gases.

Circle Graph To display data as parts of a whole, you might use a circle graph. A circle graph is a circle divided into sections that represent the relative size of each piece of data. The entire circle represents 100%, half represents 50%, and so on.

Example Air is made up of 78% nitrogen, 21% oxygen, and 1% other gases. Display the composition of air in a circle graph.

Step 1 Multiply each percent by 360° and divide by 100 to find the angle of each section in the circle.

$$78\% \times \frac{360°}{100} = 280.8°$$

$$21\% \times \frac{360°}{100} = 75.6°$$

$$1\% \times \frac{360°}{100} = 3.6°$$

Step 2 Use a compass to draw a circle and to mark the center of the circle. Draw a straight line from the center to the edge of the circle.

Step 3 Use a protractor and the angles you calculated to divide the circle into parts. Place the center of the protractor over the center of the circle and line the base of the protractor over the straight line.

Practice Problem Draw a circle graph to represent the amount of aluminum collected during the week shown in the bar graph to the left.

Physical Science Reference Tables

Standard Units

Symbol	Name	Quantity
m	meter	length
kg	kilogram	mass
Pa	pascal	pressure
K	kelvin	temperature
mol	mole	amount of a substance
J	joule	energy, work, quantity of heat
s	second	time
C	coulomb	electric charge
V	volt	electric potential
A	ampere	electric current
Ω	ohm	resistance

Physical Constants and Conversion Factors

Acceleration due to gravity	g	9.8 m/s/s or m/s^2
Avogadro's Number	N_A	6.02×10^{23} particles per mole
Electron charge	e	1.6×10^{-19} C
Electron rest mass	m_e	9.11×10^{-31} kg
Gravitation constant	G	6.67×10^{-11} N $\times$ m^2/kg^2
Mass-energy relationship		1 u (amu) = 9.3×10^2 MeV
Speed of light in a vacuum	c	3.00×10^8 m/s
Speed of sound at STP		331 m/s
Standard Pressure		1 atmosphere
		101.3 kPa
		760 Torr or mmHg
		14.7 lb/in.^{2}

Wavelengths of Light in a Vacuum

Violet	$4.0 - 4.2 \times 10^{-7}$ m
Blue	$4.2 - 4.9 \times 10^{-7}$ m
Green	$4.9 - 5.7 \times 10^{-7}$ m
Yellow	$5.7 - 5.9 \times 10^{-7}$ m
Orange	$5.9 - 6.5 \times 10^{-7}$ m
Red	$6.5 - 7.0 \times 10^{-7}$ m

The Index of Refraction for Common Substances
($\lambda = 5.9 \times 10^{-7}$ m)

Air	1.00
Alcohol	1.36
Canada Balsam	1.53
Corn Oil	1.47
Diamond	2.42
Glass, Crown	1.52
Glass, Flint	1.61
Glycerol	1.47
Lucite	1.50
Quartz, Fused	1.46
Water	1.33

Heat Constants

	Specific Heat (average) (kJ/kg $\times$ °C) (J/g $\times$ °C)	Melting Point (°C)	Boiling Point (°C)	Heat of Fusion (kJ/kg) (J/g)	Heat of Vaporization (kJ/kg) (J/g)
Alcohol (ethyl)	2.43 (liq.)	−117	79	109	855
Aluminum	0.90 (sol.)	660	2467	396	10500
Ammonia	4.71 (liq.)	−78	−33	332	1370
Copper	0.39 (sol.)	1083	2567	205	4790
Iron	0.45 (sol.)	1535	2750	267	6290
Lead	0.13 (sol.)	328	1740	25	866
Mercury	0.14 (liq.)	−39	357	11	295
Platinum	0.13 (sol.)	1772	3827	101	229
Silver	0.24 (sol.)	962	2212	105	2370
Tungsten	0.13 (sol.)	3410	5660	192	4350
Water (solid)	2.05 (sol.)	0	–	334	–
Water (liquid)	4.18 (liq.)	–	100	–	–
Water (vapor)	2.01 (gas)	–	–	–	2260
Zinc	0.39 (sol.)	420	907	113	1770

Reference Handbooks

Standard Units

Heat Constants

$^4_2$He (α particle) Helium nucleus emission

$^0_{-1}$e (β particle) electron emission

Reference Handbooks

PERIODIC TABLE OF THE ELEMENTS

Columns of elements are called groups. Elements in the same group have similar chemical properties.

- Gas
- Liquid
- Solid
- Synthetic

Element — Hydrogen
Atomic number — 1
Symbol — H
Atomic mass — 1.008
State of matter

The first three symbols tell you the state of matter of the element at room temperature. The fourth symbol identifies elements that are not present in significant amounts on Earth. Useful amounts are made synthetically.

Group	1	2	3	4	5	6	7	8	9
1	Hydrogen 1 H 1.008								
2	Lithium 3 Li 6.941	Beryllium 4 Be 9.012							
3	Sodium 11 Na 22.990	Magnesium 12 Mg 24.305							
4	Potassium 19 K 39.098	Calcium 20 Ca 40.078	Scandium 21 Sc 44.956	Titanium 22 Ti 47.867	Vanadium 23 V 50.942	Chromium 24 Cr 51.996	Manganese 25 Mn 54.938	Iron 26 Fe 55.845	Cobalt 27 Co 58.933
5	Rubidium 37 Rb 85.468	Strontium 38 Sr 87.62	Yttrium 39 Y 88.906	Zirconium 40 Zr 91.224	Niobium 41 Nb 92.906	Molybdenum 42 Mo 95.94	Technetium 43 Tc (98)	Ruthenium 44 Ru 101.07	Rhodium 45 Rh 102.906
6	Cesium 55 Cs 132.905	Barium 56 Ba 137.327	Lanthanum 57 La 138.906	Hafnium 72 Hf 178.49	Tantalum 73 Ta 180.948	Tungsten 74 W 183.84	Rhenium 75 Re 186.207	Osmium 76 Os 190.23	Iridium 77 Ir 192.217
7	Francium 87 Fr (223)	Radium 88 Ra (226)	Actinium 89 Ac (227)	Rutherfordium 104 Rf (261)	Dubnium 105 Db (262)	Seaborgium 106 Sg (266)	Bohrium 107 Bh (264)	Hassium 108 Hs (277)	Meitnerium 109 Mt (268)

The number in parentheses is the mass number of the longest-lived isotope for that element.

Rows of elements are called periods. Atomic number increases across a period.

The arrow shows where these elements would fit into the periodic table. They are moved to the bottom of the table to save space.

Lanthanide series	Cerium 58 Ce 140.116	Praseodymium 59 Pr 140.908	Neodymium 60 Nd 144.24	Promethium 61 Pm (145)	Samarium 62 Sm 150.36
Actinide series	Thorium 90 Th 232.038	Protactinium 91 Pa 231.036	Uranium 92 U 238.029	Neptunium 93 Np (237)	Plutonium 94 Pu (244)

Reference Handbooks

Metal
Metalloid
Nonmetal

The color of an element's block tells you if the element is a metal, nonmetal, or metalloid.

Science Online
Visit bookm.msscience.com for updates to the periodic table.

	13	14	15	16	17	18
						Helium 2 **He** 4.003
	Boron 5 **B** 10.811	Carbon 6 **C** 12.011	Nitrogen 7 **N** 14.007	Oxygen 8 **O** 15.999	Fluorine 9 **F** 18.998	Neon 10 **Ne** 20.180

10	11	12	13	14	15	16	17	18
			Aluminum 13 **Al** 26.982	Silicon 14 **Si** 28.086	Phosphorus 15 **P** 30.974	Sulfur 16 **S** 32.065	Chlorine 17 **Cl** 35.453	Argon 18 **Ar** 39.948
Nickel 28 **Ni** 58.693	Copper 29 **Cu** 63.546	Zinc 30 **Zn** 65.409	Gallium 31 **Ga** 69.723	Germanium 32 **Ge** 72.64	Arsenic 33 **As** 74.922	Selenium 34 **Se** 78.96	Bromine 35 **Br** 79.904	Krypton 36 **Kr** 83.798
Palladium 46 **Pd** 106.42	Silver 47 **Ag** 107.868	Cadmium 48 **Cd** 112.411	Indium 49 **In** 114.818	Tin 50 **Sn** 118.710	Antimony 51 **Sb** 121.760	Tellurium 52 **Te** 127.60	Iodine 53 **I** 126.904	Xenon 54 **Xe** 131.293
Platinum 78 **Pt** 195.078	Gold 79 **Au** 196.967	Mercury 80 **Hg** 200.59	Thallium 81 **Tl** 204.383	Lead 82 **Pb** 207.2	Bismuth 83 **Bi** 208.980	Polonium 84 **Po** (209)	Astatine 85 **At** (210)	Radon 86 **Rn** (222)
Darmstadtium 110 **Ds** (281)	Unununium * 111 **Uuu** (272)	Ununbium * 112 **Uub** (285)		Ununquadium * 114 **Uuq** (289)		** 116		** 118

* The names and symbols for elements 111–114 are temporary. Final names will be selected when the elements' discoveries are verified.

** Elements 116 and 118 were thought to have been created. The claim was retracted because the experimental results could not be repeated.

Europium 63 **Eu** 151.964	Gadolinium 64 **Gd** 157.25	Terbium 65 **Tb** 158.925	Dysprosium 66 **Dy** 162.500	Holmium 67 **Ho** 164.930	Erbium 68 **Er** 167.259	Thulium 69 **Tm** 168.934	Ytterbium 70 **Yb** 173.04	Lutetium 71 **Lu** 174.967
Americium 95 **Am** (243)	Curium 96 **Cm** (247)	Berkelium 97 **Bk** (247)	Californium 98 **Cf** (251)	Einsteinium 99 **Es** (252)	Fermium 100 **Fm** (257)	Mendelevium 101 **Md** (258)	Nobelium 102 **No** (259)	Lawrencium 103 **Lr** (262)

Glossary/Glosario

Cómo usar el glosario en español:
1. Busca el término en inglés que desees encontrar.
2. El término en español, junto con la definición, se encuentran en la columna de la derecha.

Pronunciation Key

Use the following key to help you sound out words in the glossary.

a	back (BAK)		ew	food (FEWD)
ay	day (DAY)		yoo	pure (PYOOR)
ah	father (FAH thur)		yew	few (FYEW)
ow	flower (FLOW ur)		uh	comma (CAH muh)
ar	car (CAR)		u (+ con)	rub (RUB)
e	less (LES)		sh	shelf (SHELF)
ee	leaf (LEEF)		ch	nature (NAY chur)
ih	trip (TRIHP)		g	gift (GIHFT)
i (i + con + e)	idea (i DEE uh)		j	gem (JEM)
oh	go (GOH)		ing	sing (SING)
aw	soft (SAWFT)		zh	vision (VIH zhun)
or	orbit (OR buht)		k	cake (KAYK)
oy	coin (COYN)		s	seed, cent (SEED, SENT)
oo	foot (FOOT)		z	zone, raise (ZOHN, RAYZ)

English — A — Español

acceleration: equals the change in velocity divided by the time for the change to take place; occurs when an object speeds up, slows down, or turns. (p. 14)

alternative resource: new renewable or inexhaustible energy source; includes solar energy, wind, and geothermal energy. (p. 143)

Archimedes' principle: states that the buoyant force on an object equals the weight of the fluid displaced by the object. (p. 77)

average speed: equals the total distance traveled divided by the total time taken to travel the distance. (p. 11)

aceleración: es igual al cambio de velocidad dividido por el tiempo que toma en realizarse dicho cambio; sucede cuando un objeto aumenta su velocidad, la disminuye o gira. (p. 14)

recurso alternativo: nueva fuente de energía renovable o inagotable; incluye energía solar, eólica y geotérmica. (p. 143)

principio de Arquímedes: establece que la fuerza de flotación de un objeto es igual al peso del fluido desplazado por dicho objeto. (p. 77)

velocidad promedio: es igual al total de la distancia recorrida dividida por el tiempo total necesario para recorrer dicha distancia. (p. 11)

B

balanced forces: two or more forces whose effects cancel each other out and do not change the motion of an object. (p. 37)

Bernoulli's principle: states that when the speed of a fluid increases, the pressure exerted by the fluid decreases. (p. 85)

buoyant force: upward force exerted by a fluid on any object placed in the fluid. (p. 74)

fuerzas balanceadas: dos o más fuerzas cuyos efectos se cancelan mutuamente sin cambiar el movimiento de un objeto. (p. 37)

principio de Bernoulli: establece que cuando se incrementa la velocidad de un fluido, disminuye la presión ejercida por el mismo. (p. 85)

fuerza de flotación: fuerza ascendente ejercida por un fluido sobre cualquier objeto colocado en dicho fluido. (p. 74)

Glossary/Glosario

C

center of mass: point in a object that moves as if all of the object's mass were concentrated at that point. (p. 48)

chemical energy: energy stored in chemical bonds. (p. 129)

compound machine: machine made up of a combination of two or more simple machines. (p. 109)

conduction: transfer of thermal energy by direct contact; occurs when energy is transferred by collisions between particles. (p. 163)

conductor: material that transfers heat easily. (p. 165)

convection: transfer of thermal energy by the movement of particles from one place to another in a gas or liquid. (p. 164)

centro de masa: punto en un objeto que se mueve como si toda la masa del objeto estuviera concentrada en ese punto. (p. 48)

energía química: energía almacenada en enlaces químicos. (p. 129)

máquina compuesta: máquina compuesta por la combinación de dos o más máquinas. (p. 109)

conducción: transferencia de energía térmica por contacto directo; se produce cuando la energía se transfiere mediante colisiones entre las partículas. (p. 163)

conductor: material que transfiere calor fácilmente. (p. 165)

convección: transferencia de energía térmica por el movimiento de partículas de un sitio a otro en un líquido o un gas. (p. 164)

D

density: mass of an object divided by its volume. (p. 78)

densidad: masa de un objeto dividida por su volumen. (p. 78)

E

efficiency: equals the output work divided by the input work; expressed as a percentage. (p. 107)

electrical energy: energy carried by electric current. (p. 130)

energy: the ability to cause change. (p. 126)

eficiencia: equivale al trabajo aplicado dividido el trabajo generado y se expresa en porcentaje. (p. 107)

energía eléctrica: energía transportada por corriente eléctrica. (p. 130)

energía: capacidad de producir cambios. (p. 126)

F

fluid: a substance that has no definite shape and can flow. (p. 69)

force: a push or a pull. (p. 36)

friction: force that acts to oppose sliding between two surfaces that are touching. (p. 38)

fluido: sustancia que no tiene forma definida y que puede fluir. (p. 69)

fuerza: presión o tracción. (p. 36)

fricción: fuerza que actúa para oponerse al deslizamiento entre dos superficies que se tocan. (p. 38)

G

generator: device that transforms kinetic energy into electrical energy. (p. 136)

generador: dispositivo que transforma la energía cinética en energía eléctrica. (p. 136)

Glossary/Glosario

heat/mass   **calor/masa**

H

heat: thermal energy transferred from a warmer object to a cooler object. (p. 162)

heat engine: device that converts thermal energy into mechanical energy. (p. 169)

hydraulic system: uses a fluid to increase an applied force. (p. 83)

calor: energía térmica transferida de un objeto con más calor a uno con menos calor. (p. 162)

motor de calor: motor que transforma la energía térmica en energía mecánica. (p. 169)

sistema hidráulico: usa un fluido para incrementar una fuerza aplicada. (p. 83)

I

inclined plane: simple machine that is a flat surface, sloped surface, or ramp. (p. 109)

inertia: tendency of an object to resist a change in its motion. (p. 19)

inexhaustible resource: energy source that can't be used up by humans. (p. 143)

input force: force exerted on a machine. (p. 104)

instantaneous speed: the speed of an object at one instant of time. (p. 11)

internal combustion engine: heat engine in which fuel is burned in a combustion chamber inside the engine. (p. 170)

plano inclinado: máquina simple que consiste en una superficie plana, inclinada, o una rampa. (p. 109)

inercia: tendencia de un objeto a resistirse a un cambio de movimiento. (p. 19)

recurso inagotable: fuente de energía que no puede ser agotada por los seres humanos. (p. 143)

fuerza aplicada: fuerza que se ejerce sobre una máquina. (p. 104)

velocidad instantánea: la velocidad de un objeto en un instante de tiempo. (p. 11)

motor de combustión interna: motor de calor en el cual el combustible es quemado en una cámara de combustión dentro del motor. (p. 170)

K

kinetic energy: energy an object has due to its motion. (p. 127)

energía cinética: energía que posee un objeto debido a su movimiento. (p. 127)

L

law of conservation of energy: states that energy can change its form but is never created or destroyed. (p. 132)

law of conservation of momentum: states that the total momentum of objects that collide with each other is the same before and after the collision. (p. 21)

lever: simple machine consisting of a rigid rod or plank that pivots or rotates about a fixed point called the fulcrum. (p. 112)

ley de la conservación de la energía: establece que la energía puede cambiar de forma pero nunca puede ser creada ni destruida. (p. 132)

ley de conservación de momento: establece que el momento total de los objetos que chocan entre sí es el mismo antes y después de la colisión. (p. 21)

palanca: máquina simple que consiste en una barra rígida que puede girar sobre un punto fijo llamado punto de apoyo. (p. 112)

M

mass: amount of matter in an object. (p. 19)

masa: cantidad de materia en un objeto. (p. 19)

Glossary/Glosario

mechanical advantage/power — **ventaja mecánica/potencia**

mechanical advantage: number of times the input force is multiplied by a machine; equal to the output force divided by the input force. (p. 105)

momentum: a measure of how difficult it is to stop a moving object; equals the product of mass and velocity. (p. 20)

ventaja mecánica: número de veces que la fuerza aplicada es multiplicada por una máquina; equivale a la fuerza producida dividida por la fuerza aplicada. (p. 105)

momento: medida de la dificultad para detener un objeto en movimiento; es igual al producto de la masa por la velocidad. (p. 20)

N

net force: combination of all forces acting on an object. (p. 37)

Newton's first law of motion: states that if the net force acting on an object is zero, the object will remain at rest or move in a straight line with a constant speed. (p. 38)

Newton's second law of motion: states that an object acted upon by a net force will accelerate in the direction of the force, and that the acceleration equals the net force divided by the object's mass. (p. 42)

Newton's third law of motion: states that forces always act in equal but opposite pairs. (p. 49)

nonrenewable resource: energy resource that is used up much faster than it can be replaced. (p. 140)

nuclear energy: energy contained in atomic nuclei. (p. 130)

fuerza neta: la combinación de todas las fuerzas que actúan sobre un objeto. (p. 37)

primera ley de movimiento de Newton: establece que si la fuerza neta que actúa sobre un objeto es igual a cero, el objeto se mantendrá en reposo o se moverá en línea recta a una velocidad constante. (p. 38)

segunda ley de movimiento de Newton: establece que si una fuerza neta se ejerce sobre un objeto, éste se acelerará en la dirección de la fuerza y la aceleración es igual a la fuerza neta dividida por la masa del objeto. (p. 42)

tercera ley de movimiento de Newton: establece que las fuerzas siempre actúan en pares iguales pero opuestos. (p. 49)

recurso no renovable: recurso energético que se agota mucho más rápidamente de lo que puede ser reemplazado. (p. 140)

energía nuclear: energía contenida en los núcleos de los átomos. (p. 130)

O

output force: force exerted by a machine. (p. 104)

fuerza generada: fuerza producida por una máquina. (p. 104)

P

Pascal's principle: states that the pressure applied to a confined fluid is transmitted equally throughout the fluid. (p. 83)

photovoltaic: device that transforms radiant energy directly into electrical energy. (p. 144)

potential energy: energy stored in an object due to its position. (p. 128)

power: rate at which work is done; equal to the work done divided by the time it takes to do the work; measured in watts (W). (p. 101)

principio de Pascal: establece que la presión aplicada a un fluido encerrado se transmite de manera uniforme a través de todo el fluido. (p. 83)

fotovoltaico: dispositivo que transforma la energía radiante directamente en energía eléctrica. (p. 144)

energía potencial: energía almacenada en un objeto debido a su posición. (p. 128)

potencia: velocidad a la que se realiza un trabajo y que equivale al trabajo realizado dividido por el tiempo que toma realizar el trabajo; se mide en vatios (W). (p. 101)

Glossary/Glosario

pressure/unbalanced forces   **presión/fuerzas no balanceadas**

pressure: amount of force applied per unit area on an object's surface; SI unit is the Pascal (Pa). (p. 66)

pulley: simple machine made from a grooved wheel with a rope or cable wrapped around the groove. (p. 114)

presión: cantidad de fuerza aplicada por unidad de área sobre la superficie de un objeto; la unidad internacional SI es el Pascal (Pa). (p. 66)

polea: máquina simple que consiste en una rueda acanalada con una cuerda o cable que corre alrededor del canal. (p. 114)

R

radiant energy: energy carried by light. (p. 129)

radiation: transfer of energy by electromagnetic waves. (p. 163)

renewable resource: energy resource that is replenished continually. (p. 142)

energía radiante: energía transportada por la luz. (p. 129)

radiación: transferencia de energía mediante ondas electromagnéticas. (p. 163)

recurso renovable: recurso energético regenerado continuamente. (p. 142)

S

screw: simple machine that is an inclined plane wrapped around a cylinder or post. (p. 111)

simple machine: a machine that does work with only one movement; includes the inclined plane, wedge, screw, lever, wheel and axle, and pulley. (p. 109)

specific heat: amount of heat needed to raise the temperature of 1 kg of a substance by 1°C. (p. 166)

speed: equals the distance traveled divided by the time it takes to travel that distance. (p. 10)

tornillo: máquina simple que consiste en un plano inclinado envuelto en espiral alrededor de un cilindro o poste. (p. 111)

máquina simple: máquina que ejecuta el trabajo con un solo movimiento; incluye el plano inclinado, la palanca, el tornillo, la rueda y el eje y la polea. (p. 109)

calor específico: cantidad de calor necesario para elevar la temperatura de 1 kilogramo de una sustancia en 1 grado centígrado. (p. 166)

rapidez: equivale a dividir la distancia recorrida por el tiempo que toma recorrer dicha distancia. (p. 10)

T

temperature: a measure of the average value of the kinetic energy of the particles in a material. (p. 158)

thermal energy: energy that all objects have that increases as the object's temperature increases; the sum of the kinetic and potential energy of the particles in a material. (pp. 128, 161)

thermal pollution: increase in temperature of a natural body of water; caused by adding warmer water. (p. 167)

turbine: set of steam-powered fan blades that spins a generator at a power plant. (p. 136)

temperatura: medida del valor promedio de energía cinética de las partículas en un material. (p. 158)

energía térmica: energía que poseen todos los objetos y que aumenta al aumentar la temperatura de éstos; la suma de la energía cinética y potencial de las partículas en un material. (pp. 128, 161)

polución térmica: incremento de la temperatura de una masa natural de agua producido al agregarle agua a mayor temperatura. (p. 167)

turbina: conjunto de aspas de ventilador impulsadas por vapor que hacen girar a un generador en una planta de energía eléctrica. (p. 136)

U

unbalanced forces: two or more forces acting on an object that do not cancel, and cause the object to accelerate. (p. 37)

fuerzas no balanceadas: dos o más fuerzas que actúan sobre un objeto sin anularse y que hacen que el objeto se acelere. (p. 37)

V

velocity: speed and direction of a moving object. (p. 13)

velocidad: rapidez y dirección de un objeto en movimiento. (p. 13)

W

wedge: simple machine consisting of an inclined plane that moves; can have one or two sloping sides. (p. 110)

weight: gravitational force between an object and Earth. (p. 43)

wheel and axle: simple machine made from two circular objects of different sizes that are attached and rotate together. (p. 112)

work: is done when a force exerted on an object causes that object to move some distance; equal to force times distance; measured in joules (J). (p. 98)

cuña: máquina simple que consiste en un plano inclinado que se mueve; puede tener uno o dos lados inclinados. (p. 110)

peso: fuerza gravitacional entre un objeto y la Tierra. (p. 43)

rueda y eje: máquina simple compuesta por dos objetos circulares de diferentes tamaños que están interconectados y giran. (p. 112)

trabajo: se realiza cuando la fuerza ejercida sobre un objeto hace que el objeto se mueva determinada distancia; es igual a la fuerza multiplicada por la distancia y se mide en julios (J). (p. 98)

Index

Italic numbers = illustration/photo **Bold numbers = vocabulary term**
lab = indicates a page on which the entry is used in a lab
act = indicates a page on which the entry is used in an activity

A

Acceleration, 14–18; calculating, 15–16, 16 *act*, 45, 45 *act*; equation for, 16; and force, *42, 42–43*, 46; graph of, 18, *18*; modeling, 17, 17 *lab*; and motion, 14–15; negative, 17, *17*; positive, 17; and speed, *14, 14–15*; unit of measurement with, 43; and velocity, *14, 14–15*, **15**
Action and reaction, 49–52, *50, 52*
Activities, Applying Math, 10, 16, 20, 45, 67, 100, 101, 105, 107, 160; Applying Science, 78, 142; Integrate Astronomy, 43, 69; Integrate Career, 78; Integrate Earth Science, 140; Integrate History, 107; Integrate Life Science, 10, 20, 37, 50, 87, 107, 111, 133, 135, 166, 167; Science Online, 8, 12, 39, 50, 67, 84, 102, 105, 132, 142, 170; Standardized Test Practice, 32–33, 62–63, 94–95, 122–123, 154–155, 180–181
Air conditioners, 173
Airplanes, flight of, 86, *86, 87*
Air pressure, 65 *lab*. See also Atmospheric pressure
Air resistance, 47
Alternative resources, 143 *lab*, 143–145, *144, 145*
Animal(s), effect of momentum on motion of, 20; insulation of, 166; speed of, 10
Applying Math, Acceleration of a Bus, 16; Acceleration of a Car, 45; Calculating Efficiency, 107; Calculating Mechanical Advantage, 105; Calculating Power, 101; Calculating Pressure, 67; Calculating Work, 100; Chapter Review, 31, 93, 121, 153, 179; Section Review, 13, 18, 24, 48, 54, 73, 80, 87, 102, 108, 115, 147, 161; Converting to Celsius, 160; Momentum of a Bicycle, 20; Speed of a Swimmer, 10
Applying Science, Is energy consumption outpacing production?, 142; Layering Liquids, 78
Applying Skills, 41, 130, 137, 167, 173
Archimedes' principle, 77, *77*–80, *80*
Area, and pressure, 68, *68*
Artificial body parts, 118, *118*
Atmospheric pressure, 65 *lab*, **71,** 71–73, *72, 73,* 88–89 *lab*
Automobiles, air bags in, 58, *58*; hybrid, 133, *133,* 150, *150*; internal combusion engines in, 170, *170,* 170 *act, 171;* safety in, 26–27 *lab*, 58, *58*
Average speed, 11, *11,* 11 *lab*
Axle. *See* Wheel and axle

B

Balaban, John, 90
Balanced forces, 37, *37*
Balloon races, 55 *lab*
Barometer, 73, *73,* 88–89 *lab*
Barometric pressure, 88–89 *lab*. *See also* Atmospheric pressure
Bernoulli, Daniel, 85
Bernoulli's principle, 85, *85,* 85 *lab*
Bicycles, 109, *109*
Biomechanics, 37
Bionics, 118
Birds, flight adaptations of, 87; how birds fly, 50 *act*
Black holes, 43
Boats, reason for floating, 76, *76,* 80, *80*
Body parts, artificial, 118, *118*
Body temperature, 107, 135
Boomerangs, 28, *28*
Building materials, insulators, 166, *166*
Buoyant force, *74,* **74**–81; and Archimedes' principle, 77, 77–80, *80;* cause of, 74–75; changing, 76, 76–77, *77;* and depth, 77, *77;* measuring, 81 *lab;* and shape, 76, *76;* and unbalanced pressure, 75, *75*

C

Carnivores, 111
Car safety testing, 26–27 *lab*
Celsius scale, 159, 159–160, 160 *act*
Chemical energy, 129, *129,* 133, *134*
Chemical reactions, 107
Circular motion, 46–47, *47*
Cities, heat in, 176, *176*
Coal, 140, *140*
Collisions, 7 *lab,* **21,** 21–24, *22, 23, 24,* 25 *lab,* 26–27 *lab*
Communicating Your Data, 25, 27, 55, 57, 81, 89, 103, 117, 138, 149, 168, 175
Compound machines, 109, *109*
Compressor, 172, *172*
Conduction, 163, *163*
Conductor, 165
Conservation, of energy, **132,** 139, 147, 169; of momentum, *21,* 21–24, *22, 23*
Constant speed, 11, *11*
Convection, 164–165, *165,* 165 *lab*
Coolant, 172, *172,* 173, *173*
Cooling, 168 *lab*
Crankshaft, 170
Cylinders, 170, *171*

Index

D

Data Source, 88, 148
Density, 78–79; calculating, 78 *act*; equation for, 78; and floating, 79, *79*; and sinking, 79, *79*
Depth, and buoyant force, 77, *77*; and pressure, 70
Design Your Own, Car Safety Testing, 26–27; Comparing Thermal Insulators, 174–175; Modeling Motion in Two Directions, 56–57; Pulley Power, 116–117
Direction, changing, 106, *106*; of force, 99, *99*, 106, *106*
Displacement, and distance, 9, *9*
Distance, changing, 106, *106*; and displacement, 9, *9*; and work, 100, 106, *106*
Distance-time graph, 12, *12*

E

Efficiency, 107–108, *108*; calculating, 107 *act*; equation for, 107; and friction, 108
Electrical energy, 130, 135, *135, 136,* 136–137, *137,* 169
Electricity, consumption of, 142, 142 *act,* 148–149 *lab*; generating, 136, 136–137, *137,* 142 *act, 143*; power failure, 150
Elevation, and pressure, 72
Energy, 124–150, 125 *lab*, **126,** *126,* 132 *act*; alternative sources of, 143 *lab,* 143–145, *144, 145;* chemical, **129,** *129,* 133, *134;* conservation of, **132,** 139, 147, 169; consumption of, 142, 148–149 *lab;* electrical, **130,** 135, *135, 136,* 136–137, *137,* 169; forms of, *128,* 128–130, *129, 130,* 169; from fossil fuels, 140, *140;* geothermal, 144–145, *145;* kinetic, **127,** *127,* 132, *132,* 133, *134;* mechanical, 169, *169, 171;* of motion, 127, *127;* nuclear, **130,** *130,* 141, *141,* 169; potential, **128,** *128,* 132, *132;* and power, 102;

radiant, **129,** *129,* 169; solar, 143–144, *144, 145,* 150, *150;* sources of, 137, 139–147, 148–149 *lab;* storing, 136, *136;* thermal, **128,** *128,* 135–137, *136.* See Thermal energy; tidal, 146, *146;* transfer of, 126; using, 139; wind, 145, *145;* and work, 102
Energy transformations, *131,* 131–137, 138 *lab;* analyzing, 133 *lab;* chemical energy, 133, *134;* efficiency of, *133;* electrical energy, 135, *135;* between kinetic and potential energy, 132, *132;* thermal energy, 135–136, *136;* tracking, 131
Engines, 169, **169**–171, *170,* 170 *act, 171*
Equation(s), acceleration, 16; for density, 78; for efficiency, 107; for mechanical advantage, 105; one-step, 67 *act,* 100 *act,* 101, 105 *act,* 107 *act;* for power, 101; simple, 45 *act,* 160; for work, 100
Etna, Mount (Italy), *124*
Exhaust valve, *171*
Expansion, thermal, 159, *159*

F

Fahrenheit scale, *159,* 159–160, 160 *act*
First-class lever, *113*
Fixed pulleys, 114, *115*
Flight, 50 *act,* 85–87, *86, 87*
Floating, 75, *75.* See also Buoyant force; and Archimedes' principle, 80, *80;* and boats, 76, *76,* 80, *80;* and density, 79, *79*
Fluid(s), 69, **69**–73, 82–87; pressure in, 69–71, *70, 71,* 82, *82,* 84, *84*
Fluid forces, 82–87; and Bernoulli's principle, 85, *85,* 85 *lab;* and hydraulic systems, 83, 83–84, 84 *act;* increasing, 84; and Pascal's principle, 83, *83;* using, 82, *82*
Foldables, 7, 35, 65, 97, 125, 157
Footprints, interpreting, 68 *lab*
Force(s), 36, **36**–37. See Buoyant force; Fluid forces; Pressure;

and Acceleration, 42, 42–43, 46; action and reaction, 49–52, *50, 52;* balanced, **37,** *37;* changing, 105; combining, 37; comparing, 97 *lab;* direction of, 99, *99,* 106, *106;* effects of, 35 *lab;* input, **104,** *104;* net, **37,** 46; output, **104,** *104;* unbalanced, **37;** unit of measurement with, 43; and work, 97 *lab,* 99, *99,* 103 *lab,* 105
Force pairs, measuring, 53 *lab*
Fossil fuels, as source of energy, 140, *140*
Four-stroke cycle, 170, *171*
Free fall, 53, *53,* 54
Friction, 38, **38**–41, 40 *lab,* 108, *108;* rolling, 41, *41;* sliding, 39, 40, *40, 41,* 45, 48; static, 40
Fulcrum, 112, *113*

G

Galilei, Galileo, 38, 39 *act*
Gasoline, 139, 140
Gasoline engines, 133, *133*
Generator, 136, *136*
Geothermal energy, 144–145, *145*
Graph(s), of accelerated motion, 18, *18;* distance-time, 12, *12;* of motion, 12, *12;* speed-time, 18, *18*
Gravity, 43–44; and air resistance, 47; and motion, 43, 46, *46,* 47, *47*

H

Heat, 162–168; conduction of, 163, *163;* convection of, *164,* 164–165, *165,* 165 *lab;* radiation of, 163; specific, **166;** and thermal energy, 162–165; transfer of, 162, *162*
Heat engines, 169, **169**–171, *170,* 170 *act, 171*
Heating, 168 *lab*
Heat island, 176, *176*
Heat pumps, 145, 173, *173*
Herbivore(s), 111

Index

Hybrid cars, 133, *133*, 150, *150*
Hydraulic systems, 83, *83*–84, 84 *act*
Hydroelectric power, 142, *143*

I

Inclined plane, 109–111, *110*
Inertia, 19, *19*
Inexhaustible resources, 143
Input force, 104, *104*
Instantaneous speed, 11, *11*
Insulators, 166, *166*, 174–175 *lab*
Integrate Astronomy, black holes, 43; plasma, 69
Integrate Career, Naval architect, 78
Integrate Earth Science, energy source origins, 140
Integrate History, James Prescott Joule, 107
Integrate Life Science, animal insulation, 166; animal speeds, 10; biomechanics, 37; bird's wings, 87; body temperature, 107; controlling body temperature, 135; effect of momentum on animal motion, 20; flight, 50; thermal pollution, 167; transforming chemical energy, 133; wedges in your body, 111
Internal combustion engines, 170, *170*, 170 *act*, *171*

J

Joule, James Prescott, 107
Journal, 6, 34, 64, 96, 124, 156

K

Kelvin scale, 160
Kilogram (kg), 43
Kinetic energy, 127; and mass, 127, *127*; and speed, 127, *127*; transforming chemical energy to, 133, *134*; transforming to and from potential energy, 132, *132*

L

Lab(s), Balloon Races, 55; Building the Pyramids, 103; Collisions, 25; Design Your Own, 26–27, 56–57, 116–117, 174–175; Hearing with Your Jaw, 138; Heating Up and Cooling Down, 168; Launch Labs, 7, 35, 65, 97, 125, 157; Measuring Buoyant Force, 81; MiniLabs, 17, 53, 68, 114, 143, 165; Try at Home MiniLabs, 11, 40, 85, 101, 133, 164; Use the Internet, 88–89, 148–149
Land speed, 12 *lab*
Launch Labs, Compare Forces, 97; Forces Exerted by Air, 65; How do forces affect a ball?, 35; Marbles and Energy, 125; Measuring Temperature, 157; Motion after a Collision, 7
Law(s), of conservation of energy, **132**, 139, 169; of conservation of momentum, 21, **21**–24, *22*, *23*; Newton's first law of motion, **38**–41, *41*; Newton's second law of motion, **42**–48, *51*; Newton's third law of motion, **49**–54, *51*, 55 *lab*
Lever, **112**, *112*, 113, *113*
Light, energy of, 129, *129*
Liquid(s), layering, 78 *act*

M

Machines, 104–117; compound, **109**, *109*; and efficiency, 107–108, *108*; and friction, 108, *108*; and mechanical advantage, 104, 104–106, 112, *112*; simple. *See* Simple machines
Map(s), 13, *13*
Marble launch, analyzing, 125 *lab*
Mass, 19, *19*; and kinetic energy, 127, *127*; unit of measurement with, 43; and weight, 44
Matter, and motion, 8

Measurement, of average speed, 11 *lab*; of buoyant force, 81 *lab*; of force pairs, 53 *lab*; of temperature, *159*, 159–160; units of, 16, 43, 101; of weight, 53, *53*; of work, 102
Mechanical advantage, 104, 104–106, **105**, 112, *112*
Mechanical energy, 169, *169*, 171
Melting, comparing rates of, 164 *lab*
MiniLabs, Building a Solar Collector, 143; Interpreting Footprints, 68; Measuring Force Pairs, 53; Modeling Acceleration, 17; Observing Convection, 165; Observing Pulleys, 114
Mining, of coal, 140
Momentum, **20**–24; calculating, 20 *act*; and collisions, 7 *lab*, 21, 21–24, *22*, *23*, *24*, 25 *lab*, 26–27 *lab*; conservation of, 21, 21–24, *22*, *23*
Motion, 6, 8 *act*, 8–13, 34–67; and acceleration, 14–15, *42*, 42–43, 46; after a collision, 7 *lab*; and air resistance, 47; and changing position, 8, 8–9; circular, 46–47, *47*; energy of, 127, *127*; and friction, *38*, 38–41; graphing, 12, *12*, 18, *18*; and gravity, 43, 46, *46*, 47, *47*; and matter, 8; modeling in two directions, 56–57 *lab*; and momentum, 20–24; Newton's first law of, **38**–41, *51*; Newton's second law of, **42**–48, *51*; Newton's third law of, **49**–54, *51*, 55 *lab*; on a ramp, 35 *lab*; relative, 9, *9*; and speed, 10–11, *11*; and work, 98, 98–99, *99*
Movable pulleys, 115, *115*
Muscle(s), transforming chemical energy to kinetic energy in, 133, *134*

N

National Geographic Visualizing, The Conservation of Momentum, 23; Energy Transformations, 134; The Four-Stroke Cycle, *171*;

Negative acceleration   **Speed**

Levers, *113*; Newton's laws in sports, *51*; Pressure at Varying Elevations, *72*
Negative acceleration, 17, *17*
Net force, 37, 46
Newton (unit of force), 43
Newton, Isaac, 38, 39 *act*
Newton's first law of motion, 38–41, *51*
Newton's second law of motion, 42–48, *51*; and air resistance, 47; and gravity, 43–44; using, 44, 44–46, *46*
Newton's third law of motion, 49–54, *51*, 55 *lab*
Nonrenewable resources, 140, *140*
Nuclear energy, 130, *130*, 141, *141*, 169

Ocean(s), energy from, 146, *146*
Oil (petroleum), 140
One-step equations, 67 *act*, 100 *act*, 101 *act*, 105 *act*, 107 *act*
Oops! Accidents in Science, What Goes Around Comes Around, 28
Orbit, of satellite, 47; weightlessness in, 54, *54*
Output force, 104, *104*

Pascal's principle, 83, *83*
Photovoltaic collector, 144, *144*
Pistons, 170, *171*
Plasma, 69
Polar bears, 166
Pollution, and fossil fuels, 140; thermal, **167**, *167*
Position, changing, *8*, 8–9
Positive acceleration, 17
Potential energy, 128, *128*, 132, *132*
Power, 101–102, 148–149 *lab*; calculating, 101, 101 *act*; and energy, 102; equation for, 101; geothermal, 145, *145*; hydroelectric, 142, *143*; of pulley, 116–117 *lab*; and work, 101 *lab*
Power failure, 150

Power plants, *136*, 136–137
Pressure, 66, **66**–73; of air, 65 *lab*; and area, 68, *68*; atmospheric, 65 *lab*, 71, 71–73, *72*, *73*, 88–89 *lab*; barometric, 88–89 *lab*; and buoyant force, 75, *75*; calculating, 67 *act*; and elevation, *72*; in fluid, 69–71, *70*, *71*, 82, *82*, 84, *84*; increase with depth, 70; and weight, 66, *66*, 68
Prostheses, 118, *118*
Pulley, 114 *lab*, **114**–117; fixed, 114, *115*; movable, 115, *115*; power of, 116–117 *lab*
Pulley system, 115, *115*, 116–117 *lab*
Pyramids, building, 97 *lab*, 103 *lab*

Radiant energy, 129, *129*, 169
Radiation, 163
Radio, 135, *135*, 138 *lab*
Radioactive wastes, 141
Reaction, and action, 49–52, *50*, *52*
Reactions, chemical, 107
Reading Check, 9, 11, 18, 39, 40, 42, 68, 71, 75, 79, 84, 86, 98, 100, 105, 107, 111, 114, 127, 129, 132, 136, 143, 144, 159, 163, 165
Reading Strategies, 8A, 36A, 66A, 98A, 126A, 158A
Real-World Questions, 25, 26, 55, 56, 81, 88, 103, 116, 138, 148, 168, 174
Refrigerators, 172, *172*
Relative motion, 9, *9*
Renewable resources, 142, *143*
Resources, alternative, 143 *lab*, 143–145, *144*, *145*; inexhaustible, **143**; nonrenewable, **140**, *140*; renewable, **142**, *143*
Rockets, balloon, 55 *lab*; launching, 52, *52*
Rolling friction, 41, *41*

Safety, and air bags, 58, *58*; in automobiles, 26–27 *lab*, 58, *58*

Satellite(s), 47
Science and Language Arts, "Hurricane" (Balaban), 90
Science and Society, Air Bag Safety, 58; Bionic People, 118; The Heat Is On, 176
Science Online, automobile engines, 170; energy and energy resources, 132, 142; Galileo and Newton, 39; historical tools, 105; how birds fly, 50; hydraulic systems, 84; land speed, 12; snowshoes, 67; studying motion, 8; James Watt, 102
Science Stats, 150
Scientific Methods, 25, 26–27, 55, 56–57, 81, 88–89, 103, 116–117, 138, 148, 148–149, 168, 174–175; Analyze Your Data, 27, 55, 57, 89, 117, 149, 175; Conclude and Apply, 25, 27, 55, 57, 81, 89, 103, 117, 138, 149, 168, 175; Follow Your Plan, 57, 89, 117, 175; Form a Hypothesis, 26, 56, 116, 148, 174; Make a Plan, 57, 88, 117, 175; Test Your Hypothesis, 27, 56, 149, 175
Screw, 111, *111*
Sea level, pressure at, *72*
Second-class lever, 113
Shape, and buoyant force, 76, *76*
Simple machines, 109–117; inclined plane, 109–111, *110*; lever, 112, *112*, 113, *113*; pulley, 114 *lab*, 114–117, *115*, 116–117 *lab*; screw, 111, *111*; wedge, *110*, 110–111, *111*; wheel and axle, 112, *112*, 114, *114*
Sinking, 75, *75*, 79, *79*. See also Buoyant force
Sliding friction, 39, 40, *40*, *41*, 45, 48
Snowshoes, 67 *act*
Solar collector, 143 *lab*, 144, *144*
Solar energy, 143–144, *144*, *145*, 150, *150*
Space shuttle, 54, *54*
Specific heat, 166
Speed, 10–11; and acceleration, 14, 14–15; of animals, 10; average, **11**, *11*, 11 *lab*; calculating, 10 *act*; constant, 11, *11*;

Index

Speed-time graph
and distance-time graphs, 12, *12*; of heating and cooling, 168 *lab*; instantaneous, **11**, *11*; and kinetic energy, 127, *127*; land, 12 *lab*; and motion, 10–11, *11*; and velocity, 13

Speed-time graph, 18, *18*

Sports, Newton's laws in, *51*

Standardized Test Practice, 32–33, 94–95, 122–123, 180–181

Static friction, 40

Study Guide, 29, 59, 91, 119, 151, 177

Sun, as plasma, 69

Technology, air conditioners, 173; airplanes, 86, *86*, *87*; barometers, 73, *73*, 88–89 *lab*; bicycle, 109, *109*; boats, 76, *76*, 80, *80*; gasoline engines, 133, *133*; generator, 136, *136*; heat pumps, 145, 173, *173*; hydraulic systems, 83, 83–84, 84 *act*; internal combustion engines, 170, *170*, 170 *act*, *171*; photovoltaic collector, 144, *144*; power plants, *136*, 136–137; pyramids, 97 *lab*, 103 *lab*; radio, 135, *135*, 138 *lab*; refrigerators, 172, *172*; rockets, 52, *52*, 55 *lab*; satellites, 47; solar collector, 143 *lab*, 144, *144*; space shuttle, 54, *54*; thermometers, 159, *159*; turbine, 136, *136*; windmill, 145, *145*

Teeth, of herbivores and carnivores, 111, *111*

Temperature, 157 *lab*, 158, **158**–160; of body, 107, 135; measuring, 159, 159–160; and thermal energy, 161, *161*

Temperature scales, Celsius, *159*, 159–160, 160 *act;* converting, 160, 170 *act;* Fahrenheit, *159*, 159–160, 160 *act;* Kelvin, 160

Thermal collector, 144, *144*

Thermal conductors, 165

Thermal energy, 128, *128*, 135–136, *136*, 156–176, **159**, *161*, 169, *169*, *171;* and heat, 162–165; and temperature, 161, *161;* transfer of, 162–164, *163*, *164*

Thermal expansion, 159, *159*

Thermal insulators, 166, *166*, 174–175 *lab*

Thermal pollution, 167, *167*

Thermometer, 159, *159*

Third-class lever, *113*

Tidal energy, 146, *146*

TIME, Science and Society, 58, 118, 176

Tools, historical, 105 *act*

Try at Home MiniLabs, Analyzing Energy Transformations, 133; Comparing Rates of Melting, 164; Measure Average Speed, 11; Observing Bernoulli's Principle, 85; Observing Friction, 40; Work and Power, 101

Turbine, 136, *136*

Unbalanced forces, 37
Uranium, 141

Use the Internet, Barometric Pressure and Weather, 88–89; Energy to Power your Life, 148–149

Velocity, 13; and acceleration, 14, 14–15, *15*; and speed, 13

Volcanoes, energy from, 124, *124*

Waste(s), radioactive, 141
Watt (W), 101
Watt, James, 101, 102 *act*
Weather, and barometric pressure, 88–89 *lab*
Wedge, *110*, **110**–111, *111*
Weight, 43–44; and mass, 44; measuring, 53, *53*; and pressure, 66, *66*, 68
Weightlessness, *53*, 53–54, *54*
Wheel and axle, 112, *112*, 114, *114*
Wind energy, 145, *145*
Windmill, 145, *145*
Wings, 86, *86*, 87, *87*
Work, 98–100; calculating, 100, 100 *act*; and distance, 100, 106, *106*; and energy, 102; equation for, 100; and force, 97 *lab*, 99, *99*, 103 *lab*, 105; measuring, 102; and mechanical advantage, 104, 104–106; and motion, 98, 98–99, *99*; and power, 101 *lab*

Credits

Magnification Key: Magnifications listed are the magnifications at which images were originally photographed.
LM–Light Microscope
SEM–Scanning Electron Microscope
TEM–Transmission Electron Microscope

Acknowledgments: Glencoe would like to acknowledge the artists and agencies who participated in illustrating this program: Absolute Science Illustration; Andrew Evansen; Argosy; Articulate Graphics; Craig Attebery represented by Frank & Jeff Lavaty; CHK America; John Edwards and Associates; Gagliano Graphics; Pedro Julio Gonzalez represented by Melissa Turk & The Artist Network; Robert Hynes represented by Mendola Ltd.; Morgan Cain & Associates; JTH Illustration; Laurie O'Keefe; Matthew Pippin represented by Beranbaum Artist's Representative; Precision Graphics; Publisher's Art; Rolin Graphics, Inc.; Wendy Smith represented by Melissa Turk & The Artist Network; Kevin Torline represented by Berendsen and Associates, Inc.; WILDlife ART; Phil Wilson represented by Cliff Knecht Artist Representative; Zoo Botanica.

Photo Credits

Cover Gunter Marx Photography/CORBIS; **i ii** Gunter Marx Photography/CORBIS; **iv** (bkgd)John Evans, (inset)Gunter Marx Photography/CORBIS; **v** (t)PhotoDisc, (b)John Evans; **vi** (l)John Evans, (r)Geoff Butler; **vii** (l)John Evans, (r)PhotoDisc; **viii** PhotoDisc; **ix** Aaron Haupt Photography; **x** Runk/Schoenberger for Grant Heilman; **xi** Billy Hustace/Stone/Getty Images; **xii** Tom & DeeAnn McCarthy/The Stock Market/CORBIS; **1** Lori Adamski Peek/Stone/Getty Images; **2** (t)Jeremy Woodhouse/Photodisc, (b)Ted Spiegel/CORBIS; **3** (t)William James Warren/CORBIS, (b)CORBIS; **5** Dominic Oldershaw; **6–7** William Dow/CORBIS; **8** Telegraph Colour Library/FPG/Getty Images; **9** Geoff Butler; **12** Richard Hutchings; **15** Runk/Schoenberger from Grant Heilman; **17** Mark Doolittle/Outside Images/Picturequest; **19** (l)Ed Bock/The Stock Market/CORBIS, (r)Will Hart/PhotoEdit, Inc.; **21** (t)Tom & DeeAnn McCarthy/The Stock Market/CORBIS, (bl)Jodi Jacobson/Peter Arnold, Inc., (br)Jules Frazier/PhotoDisc; **22** Mark Burnett; **24** Robert Brenner/PhotoEdit, Inc.; **25** Laura Sifferlin; **26 27** Icon Images; **28** Alexis Duclos/Liaison/Getty Images; **29** (l r)Rudi Von Briel/PhotoEdit, Inc., (c)PhotoDisc; **33** (l)Jodi Jacobson/Peter Arnold, Inc., (r)Runk/Schoenberger from Grant Heilman; **34–35** Wendell Metzen/Index Stock; **35** Richard Hutchings; **36** (l)Globus Brothers Studios, NYC, (r)Stock Boston; **37** Bob Daemmrich; **38** (t)Beth Wald/ImageState, (b)David Madison; **39** Rhoda Sidney/Stock Boston/PictureQuest; **41** (l)Myrleen Cate/PhotoEdit, Inc., (r)David Young-Wolff/PhotoEdit, Inc.; **42** Bob Daemmrich; **44** (t)Stone/Getty Images, (b)Myrleen Cate/PhotoEdit, Inc.; **46** David Madison; **48** Richard Megna/Fundamental Photographs; **49** Mary M. Steinbacher/PhotoEdit, Inc.; **50** (t)Betty Sederquist/Visuals Unlimited, (b)Jim Cummins/FPG/Getty Images; **51** (tl)Denis Boulanger/Allsport, (tr)Donald Miralle/Allsport, (b)Tony Freeman/PhotoEdit/PictureQuest; **52** (t)David Madison, (b)NASA; **54** NASA; **55** Richard Hutchings; **56 57** Mark Burnett; **58** (t)Tom Wright/CORBIS, (b)Didier Charre/Image Bank; **59** (tl)Philip Bailey/The Stock Market/CORBIS, (tr)Romilly Lockyer/Image Bank/Getty Images, (bl)Tony Freeman/PhotoEdit, Inc.; **63** Betty Sederquist/Visuals Unlimited; **64–65** Hughes Martin/CORBIS; **65** Matt Meadows; **66** David Young-Wolff/PhotoEdit, Inc.; **68** Runk/Schoenberger from Grant Heilman; **69** Dominic Oldershaw; **70** (t)Matt Meadows, (b)Tom Pantages; **72** (t)Bobby Model/National Geographic Image Collection, (cl)Richard Nowitz/National Geographic Image Collection, (cr)George Grall/National Geographic Image Collection, (bl)Ralph White/CORBIS, (br)CORBIS; **74** Ryan McVay/PhotoDisc; **75** CORBIS; **76** (t)Matt Meadows, (b)Vince Streano/Stone/Getty Images; **77 79** Matt Meadows; **81** John Evans; **83** KS Studios; **84** Dominic Oldershaw; **87** (t)Michael Collier/Stock Boston, (bl)George Hall/CORBIS, (br)Dean Conger/CORBIS; **88** Steve McCutcheon/Visuals Unlimited; **89** Runk/Schoenberger from Grant Heilman; **90** AP/Wide World Photos/Ray Fairall; **91** (l)D.R. & T.L. Schrichte/Stone/Getty Images, (r)CORBIS; **92** Matt Meadows; **95** Vince Streano/Stone/Getty Images; **96–97** Rich Iwasaki/Getty Images; **97** Mark Burnett; **98** Mary Kate Denny/PhotoEdit, Inc.; **106** (l)David Young-Wolff/PhotoEdit, Inc., (r)Frank Siteman/Stock Boston; **109** Duomo; **110** Robert Brenner/PhotoEdit, Inc.; **111** (t)Tom McHugh/Photo Researchers, (b)Amanita Pictures; **112** Amanita Pictures; **113** (t)Dorling Kindersley, (bl br)Bob Daemmrich; **114** (l)Wernher Krutein/Liaison Agency/Getty Images, (r)Siegfried Layda/Stone/Getty Images; **116** Tony Freeman/PhotoEdit, Inc.; **117** Aaron Haupt; **118** (t)Ed Kashi/CORBIS, (b)James Balog; **119** (l)Inc. Janeart/The Image Bank/Getty Images, (r)Ryan McVay/PhotoDisc; **123** (l)Comstock Images, (r)PhotoDisc; **124–125** Chris Knapton/Science Photo Library/Photo Researchers; **125** Matt Meadows; **126** (l c)file photo, (r)Mark Burnett; **127** (t b)Bob Daemmrich, (c)Al Tielemans/Duomo; **128** KS Studios; **129** (l r)Bob Daemmrich, (b)Andrew McClenaghan/Science Photo Library/Photo Researchers; **130** Mark Burnett/Photo Researchers; **131** Lori Adamski Peek/Stone/Getty Images; **132** Richard Hutchings; **133** Ron Kimball/Ron Kimball Photography; **134** (t)Judy Lutz, (b)Lennart Nilsson; **136 138** KS Studios; **144** (t)Jeremy Burgess/Science Photo Library/Photo Researchers, (b)John Keating/Photo Researchers; **145** Geothermal Education Office; **146** Carsand-Mosher; **147** Billy Hustace/Stone/Getty Images; **148** SuperStock; **149** Roger Ressmeyer/CORBIS; **150** (tl)Reuters NewMedia, Inc./CORBIS, (tr)PhotoDisc, (br)Dominic Oldershaw; **151** (l)Lowell Georgia/CORBIS, (r)Mark Richards/PhotoEdit, Inc.; **156–157** Peter Walton/Index Stock; **158** John Evans; **159** (t)Nancy P. Alexander/Visuals Unlimited, (b)Morton & White; **161** Tom Stack & Assoc.; **162** Doug Martin; **163** Matt Meadows; **164** Jeremy Hoare/PhotoDisc; **165** Donnie Kamin/PhotoEdit, Inc.; **166** SuperStock; **167** Colin Raw/Stone/Getty Images; **168** Aaron Haupt; **169** PhotoDisc; **170** (l)Barbara Stitzer/PhotoEdit, Inc., (c)Doug Menuez/PhotoDisc, (r)Addison Geary/Stock Boston; **172** C. Squared Studios/PhotoDisc; **174 175** Morton & White; **176** (bkgd)Chip Simons/FPG/Getty Images, (inset)Joseph Sohm/CORBIS; **177** SuperStock; **180** John Evans; **181** Michael Newman/Photo Edit, Inc.; **182** PhotoDisc; **184** Tom Pantages; **188** Michell D. Bridwell/PhotoEdit, Inc.; **189** (t)Mark Burnett, (b)Dominic Oldershaw; **190** StudiOhio; **191** Timothy Fuller; **192** Aaron Haupt; **194** KS Studios; **195** Matt Meadows; **196** Mark Burnett; **198** John Evans; **199** Amanita Pictures; **200** Bob Daemmrich; **202** Davis Barber/PhotoEdit, Inc.